Praise For N

"FIVE STARS! Deeply reflective, vividly detailed that traces a winding journey... on the life and death of a charismatic and controversial figure... Hinton's narrative style is warm and confessional, often humorous, and peppered with pop culture references, period music, and historical events. The book opens with a funeral, then moves back to trace the complex relationship of high school acquaintances... later bound by a web of competitive friendship, unspoken grudges, and romantic entanglements. Nether Land is one of the best books you'll read this year. It's truly fascinating."

—**Carol Thompson,** Readers' Favorite Reviews

"Eureka, a historian who writes like a poet. Such artful prose with honeyed metaphor, and a syntax like a conversation between old friends... a contemporary Chaucerian morality tale. I found it hard to put down, so I didn't... captivating to the last word and period."

—**Paul Harmon,** international artist and author of Inner Voices and Crossing Borders

"Equal parts illuminating and mysterious, the memoir provides a split-screen narrative of two men from the same town with the same background going on to lead vastly different lives, along with insights into the era... Fans of unsolved mysteries and personal cultural histories will enjoy Hinton's considerations of the way lives converge, diverge, and reflect each other. Nether Land is a somber, gently provocative memoir suffused with loss."

—**BookLife Reviews**

"Funny, tender, and insightful. Nether Land takes us on a journey of memory and discovery, revealing a classic American story of ambition, achievement, greed—and revenge. Kem Hinton combines witty memoir with gumshoe reporting to tell an extraordinary tale well worth reading."

—**Elaine Weiss,** author of The Woman's Hour and Spell Freedom: The Underground Schools

"Enter the world of shady dealings, Ponzi schemes, and even murder in this captivating, turn-pager memoir. You won't want to put it down."
—**Debby M. Schriver,** author of *Whispering in the Daylight*

"A thoughtful and fascinating account of the relationship between the author and one of his more intriguing and elusive classmates which may reveal as much about the narrator as about the subject."
—**Bill Carey,** author of *Fortunes, Fiddles, & Fried Chicken* and *Runaways, Coffles, & Fancy Girls*

"Just the neighborhood kids growing up? How about ingenious success funneled into a wormhole of terrible tragedy! Don't miss this page-turner."
—**Rick Glaze**, Amazon best-selling author of *Eight Pieces of Eight* and *Ralph & Murray*

"A noted architect, Kem Hinton spins a very interesting tale about the life of a former classmate, the person's strange ascension to great wealth, and his mysterious death in a faraway paradise."
—**Hal Hardin,** attorney, former judge and United States Attorney

"Part murder mystery, part memoir – hard to stop reading."
—**Phyllis Gobbell,** author of *Notorious Nashville*

"Hinton's factual account of a high school classmate's life from the outside perspective of a mere acquaintance blends elements of biography, autobiography, mystery and nostalgia that makes a compelling story, particularly for readers who remember coming of age in the 1970s."
—**Sam D. Elliott,** author of *John C. Brown of Tennessee* and member of Tennessee Historical Commission

"Both of us need to remember this time." (Discussing early versions of *Nether Land*)
—**Robert Hicks,** author of *The Widow of the South* and *A Separate Country*

NETHER LAND

High School Temptation, Insurance Haze,
Stolen Resolution, and Murder

KEM HINTON

Copyright © 2025 by Kem Hinton

All rights reserved. No part of this book may be reproduced, scanned, stored in a retrieval system or transmitted in any form by any means — electronic, mechanical, photocopy, recording, or otherwise — except for brief quotations for the purpose of review or comment, without prior permission of the author and publisher.

Printed in the United States of America

Library of Congress Control Number: 2025910148

ISBN 979-8-218-68520-1 (Print)

ISBN 979-8-218-68521-8 (eBook)

Cover/Book design by VNO Design

Cover Photograph by Steve Simonsen

TENEBO Publishing Company
Nashville, Tennessee

DEDICATION

This memoir is dedicated to my friends and fellow students at Campus School, Central High School, and the University of Tennessee, reminding all of us of the impact of one person.

Murfreesboro, Tennessee

Historic map of St. John Island　　　　　　　　　　　　　　　*USVI*

CONTENTS

1.	First Words	1
2.	Beginning Days	2
3.	Large Life at Large School	7
4.	Senior Year	17
5.	First College Days	30
6.	More at the Big Orange	49
7.	Next Adventures	59
8.	Overture to Sagas	64
9.	Trash Endeavor One	67
10.	Trash Endeavor Two	75
11.	A Past to Remember	78
12.	Making it Big in Mississippi	82
13.	Planting of the Plant	84
14.	Trashy Competition	88
15.	Promises and Science	95
16.	Into the Trash Can	98
17.	Positive Fallout	102
18.	Dead Janitor Insurance	105
19.	Reunion Justice	108
20.	The Other Ed	110
21.	Lilac Flowers	115
22.	Blooming Fortunes	118

CONTENTS

23.	Come Tumbling Down	123
24.	The Need Greed	126
25.	Tennessee Proposal	130
26.	Island of Paradise	132
27.	The Final Pitch	139
28.	Ending in Paradise	142
29.	Funeral Service	149
30.	Aftermath	153
31.	The Mirror	158
32.	Haze of Purple	165

PROLOGUE

I am intoxicated by history. It never quits pulling me backwards to a land that may have never existed. The hard task is to discover the truth, savor its sweetness and terror, constantly remember, and with wisdom make the new.

—Alfred Gardner Smith

Ed Netherland, 1954-2014

CHAPTER 1
First Words

When the funeral service ended on that overcast November day in 2014, the preacher somberly informed the congregation that the burial would be private and that the family would now leave alone. I was seated on the back pew. Just as the family moved through the center aisle, my old friend Hollywood tapped my shoulder and whispered, "I need your help with this." I had been informed by high school classmates from long ago that a fellow student, now dead and lying in the approaching cherry casket, had been the victim of a vicious crime. I'd been filled with a strange, vested interest in reconnecting with this individual, but now that would never occur. One week earlier on November 18, at a secluded location on the paradise of St. John Island, Ed Netherland had been murdered. I would soon see him slide into the abyss.

CHAPTER 2
Beginning Days

Though we were both raised in Murfreesboro, Tennessee, Edward Hutton Netherland and I lived in neighborhoods on opposite sides of this quiet, college town thirty miles south of Nashville. We were both born in 1954. I grew up in a modest house that my parents built in 1960. It was in a new subdivision called Riverview, its many homes constructed on former farmland less than a mile from the crossing where in late 1862, Union and Confederate forces bloodied the pristine waters during the epic Battle of Stones River. Ed lived on the southern side of town in a stately house on prominent Broad Street.

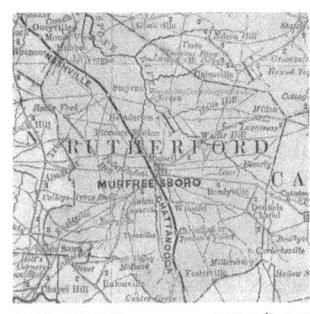

Rutherford County *TSLA/MGH*

Murfreesboro, TN *ADS*

Ed and I were never friends, and this true story came about because of my interactions with him in our high school years, occasional meetings during and after college, and several strange events that

NETHER LAND

Campus School RCH   *Bellwood School* RCH

occurred near and at the time of his death. It begins with our early years, then transitions to his fascinating and perplexing career. Ed was incredibly smart, and ten years after his mysterious and still unsolved murder, I believe his complex story should be told.

I first remember Ed in the early summer of 1965 when we both joined Boy Scout Troop 359. We were near the end of fifth grade and attending different elementary schools. Ed's folks were members of the Church of Christ while mine were at First Methodist, so our paths had never crossed before. This specific troop, one of several in our town, was composed mainly of boys like me who attended Campus School, a teaching demonstration facility for nearby Middle Tennessee State University. I heard that Ed attended Bellwood Elementary in the south part of Murfreesboro, and I wondered why he was at our first few meetings instead of with a troop on his side of town. He was stocky, and I was thin. The group was led by the exceptional scoutmaster Percy Dempsey, one who tried to teach us the Scout Laws while making sure we were excited about summer camp at

Ted LaRoche honoring Scoutmaster Dempsey

Boxwell Scout Reservation, prepared to explore the wonders of Cumberland Caverns, and eager for hikes, campouts, and other outings in nature. But at the first few gatherings, Ed seemed disinterested in these upcoming activities, perhaps preferring to spend time in sports or something else. He soon stopped attending the weekly meetings.

Life in elementary Campus School was simultaneously delightful and perplexing, with extraordinary growth between being a little kid in the first grade to a teenager in the eighth. I'd characterize mine as mostly joyful until the serious arrival of puberty threw a curve ball at me. Once consumed with making model cars, going into caves, playing baseball, and handling firecrackers, I soon encountered the two most dangerous elements that would inevitably change and sometimes inflict harm on all young boys: perfume and gasoline. But the most damaging strikeout pitch for me was the spelling of my name.

Ed certainly never had to worry about his. Mine was originally spelled Kim because my mom liked the 1901 Rudyard Kipling book about an orphaned boy in India. That name seemed okay until I was in the sixth grade. At that time of rapid maturity, my boyhood friends at Campus School, in Boy Scouts, and on my Little League baseball team received a blue box in the mail, one full of sample razors, shave cream, deodorant, and other items for young growing boys. It was an ingenious marketing ploy, and my grade school pals Buddy, John, Paul, Ned, and Duane—these among many in those elementary years that would remain friends for the rest of my life—received that blue box. I received a pink box. You likely know what was in that package: the stuff for the changing needs of maturing young girls. I was not amused, and at the time had no interest in that blossoming of the fairer sex. Thus, my name became increasingly embarrassing.

After the eighth grade in 1968, my parents arranged my first summer job as a worker on the maintenance staff at Stones River Country Club, joining three other boys whose parents, like mine, were members. We made for an odd foursome. The biggest kid in our group, Ira, who would later become All-State center of our high school's state championship football team, had that unusual first name. However, he preferred a common derivation of his middle name of Bradford and was known to all of us as "Buddy." He had a genuinely kind and fun attitude, yet if anyone teased him about that rarely-used first name, this wide-shouldered, brawny boy would have folded up the unfortunate jester and crushed him like an aluminum can.

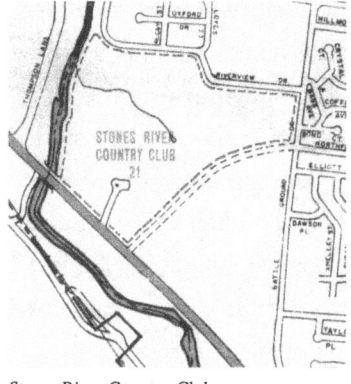
Stones River Country Club

The other large kid who would eventually become the quarterback of that same championship football team was David, and he'd receive the appropriate nickname "Hollywood" due to his brash confidence, striking blonde hair, and love of the limelight. Both of these fellow summer workers were what I would never be: tall, athletic, and super confident. The fourth boy was Jayme, and the two of us were smaller and thin with limited interest and soggy abilities in sports. As a duo, Jayme and I were often tag-teamed for specific golf course beautification tasks. During those assignments, we'd discuss politics and science. In that now-unimaginable era before the introduction of the Weed Eater, the two of us were charged on every Friday with trimming the base of all the trees on the golf course using hand clippers. It took the entire day, and only after a full weekend off would our sore mitts recover.

The four of us worked together for three summers. The job provided us $40 per week of spending cash, which burned large holes in our young pockets. Once, while our group was working on a fairway, I saw Ed in the distance, lounging near the club's tennis court. His parents also belonged to this club, and I assumed he was teaching lessons or relaxing between sets. Because my compatriot Hollywood also attended Bellwood Elementary, he and Ed were friends and waved to each other. Oddly, neither Ed nor I acknowledged each other. He seemed as distant in personality and attitude as in physical appearance, though I'm not sure why.

CHAPTER 3
Large Life at Large School

In the fall of 1968, we young whippersnappers became freshmen at the largest of only two public secondary educational facilities in the surrounding, 624-square-mile Rutherford County: Central High School. From the relatively quiet and secure settings of our separate elementary learning situations, we were thrust into an intimidating arena with more than 2,000 students, at that time the largest secondary school in Tennessee. The majority of older kids in Rutherford County attended this mammoth place that, true

Rutherford County Courthouse

to its name, was located in the smack middle of the county seat of Murfreesboro, six blocks from the paramount public square and its majestic 1859 county courthouse. Central High had been erected in 1950 on a spacious 25-acre parcel, a plot of land that had once been the location of a women's college. Its long primary façade faced picturesque East Main Street, and in 1964, when the structure was expanded on both ends, the two-floor building became an even more dominant landmark on the town's finest thoroughfare of historic residential architecture.

Central High School

Just before the start of our freshman year in the fall of 1968, officials closed Holloway High, the county's secondary school for African-American students. Its entire black student population and their teachers were transferred to the land of white ones. With that dramatic move, Central became immediately and fully integrated. Unfortunately, the forced integration strategy just north in Metropolitan Nashville & Davidson County slowly backfired. Many parents there were outraged at the use of their children as political pawns. The long-distance crosstown mixing technique, especially by busing, for racial balance and harmony instead inflicted long-term damage to the public school system as white parents helped to establish many new private and religious institutions that could avoid the intermingling of white and black children. Integration had also been required in Rutherford County, but thankfully the mixing of races in its public schools occurred peacefully. Years later, a prominent African American historian in Nashville told me, "Many of us really didn't want integration. We just wanted an end to segregation and being treated as second-class people."

At Central, different components of society, income, race, creed, religion, size, shape, talent, and intelligence were all stuffed into this 160,000-square-foot educational structure, preparing teenagers for the promises, dreams, and brutal realities of an adulthood soon to arrive. The teachers and facilities were exceptional, and only a few parents sent their kids away to private schools located in Nashville and Bell Buckle. Central's fortress of education had a wide, linear hallway that seemed to go on forever, its length lined with metal lockers and up above, photo composites of past graduating classes. At each hourly change of classes, this long passage would erupt, flooded with the energy of apprehensive, rowdy teens, each scrambling to get to their next classroom. Ed and I shared few classes together at Central, yet I recognized him occasionally in this sea of teenage humanity, its hallways crowded with so many different types of oddball characters—including me with my relentlessly-teased first

Author as freshman

Ed as freshman

name. These congested passages provided wonderful, continuously interesting interactions... and a few unpleasant moments.

In my first year at Central, I discovered that there were many other boys also dealing with gender-flexible names such as Lynn, Francis, Cameron, and Gene. At least I wasn't completely alone, although the stunning movie star Kim Novak certainly tagged that fairer-sex spelling in a most delightful, curvaceous, and permanent way. Regardless, another summer in the sun laboring at the country club's golf course was a beneficial pause from the books, giggling girls, and teasing guys at Central.

When my sophomore year started, the Apollo 11 moon landing on July 20, 1969, and the recent rock music festival at Woodstock were the primary topics discussed by happily-reunited students. Unfortunately, another summertime event had occurred on July 2 when Johnny Cash released a song previously recorded at the San Quentin State Prison in California. It was entitled "A Boy Named Sue." It was a major hit, but I never wanted to hear it. It was my story, one that I was not the least pleased with.

In January, I turned sixteen and immediately passed the driver's license exam. Life's opportunities and speed bumps would now be encountered as I cruised in my dad's white 1963 V.W. Bug. This new freedom was exhilarating, but the confusing realities of interactions between high school students would often spoil the otherwise fun times. During one change of classes that second year, I recall seeing Ed hassling a shorter student who was in my Algebra class. I was too far away to hear the specifics, but it involved unpleasant comments about the boy's older sister. Voices grew louder and pushing started. The brave kid didn't back down, and the altercation ended without a fistfight. The event stamped a definitive impression on me about the bigger, rowdy character from the other side of town.

Meanwhile, I continued to struggle along under the burden of my ill-fitting first name. Fortunately, suggestions for not only change but vast improvement would slowly arrive, and a life-altering revelation for me came inside the pages of a paperback book I purchased just as the sophomore year ended.

At home late one night after another workday with my three amigos on the country club golf course, I thumbed through the small publication containing short science fiction stories. One was titled "The Trap." It was well written, and when I turned back the pages to see the author's name, I was delivered a gift from high above. The writer was Kem Bennett. This was a eureka moment, and I was so absolutely grateful to see this revised spelling that I wanted to change his story's title to "Escape from the Trap."

I immediately convinced my dad and my initially sad mom that an official name change was required. After going through the legal process, a clerk at the state agency holding birth records would use a cheap ballpoint pen to scratch a single line through my typed first name and hurriedly etch "Kem" directly above. A desperate escape had occurred, and for the rest of my life, when asked about my unusual name and if "Kem" was short for something, I would cheerfully reply, "No. It's long for nothing."

I continued to pass Ed in the hallway at Central during our third year. We were in only a few common classes, although we both wasted time at the silenced "study hall" in the school's large and unattractive auditorium. We usually acknowledged each other's presence but nothing else. At lunchtime in Central's massive cafeteria, known for its Minus-3 Michelin star cuisine, the two of us stayed apart and munched with very different companions. Ed had unbridled gusto and was usually flirting with sweet coeds at the tables of the most popular ones, something I never attempted even with my new and improved name. From a respectable family, Ed was athletic and handsome. His two older sisters were lovely and popular, his older brother would graduate from the University of Pennsylvania's distinguished Wharton Graduate School of Business, and his father was a successful insurance businessman and lay preacher at their large Church of Christ. Popular with many, Ed possessed rock-solid self-confidence, was in the elite high school fraternity, enrolled in the advanced, mainly white classes, and immersed in athletics.

However, I knew Ed mainly as a rapscallion, a mischievous character who at any moment would pull a fast one, play a trick on an unsuspecting target, or do something conniving and sneaky. One girl who attended Sunday School with Ed recalled that he was quite unpredictable, and volunteer parents who taught the weekly religious sessions would often lament, "What has he done now?" I also knew Ed as a bully. I witnessed him harassing students, mostly younger or less athletic; if they pushed back, the result would have been an embarrassing pummeling. Occasionally he taunted female students, saying unkind things about their appearance or clothes. A sweet, tall coed told me that Ed completely demolished her emotions when she walked into a classroom clad in a jade outfit, to which he loudly commented, "Hey, it's the Jolly Green Giant!"

CHS football managers (Ed lower right)

Why did this handsome jock do these things? Ed and I both came from privileged upbringings in affluent neighborhoods, and neither had a reason to be arrogant, cruel, or abrasive to anyone.

Smart and stout, Ed loved sports, and he became a trainer and manager for Central High's football powerhouse of agility and brute strength. With other managers and the leadership of Hollywood, my former golf course co-worker and now the team's quarterback, the Central Tigers won the ultimate award for hard work, stamina, and overwhelming prowess, becoming on November 27, 1970, the Tennessee state champions. It's hard to explain how much pride that brought to every single student crammed together in the county's giant house of learning on East Main.

Football State Champions, 1970

Ed certainly deserved to be proud of his work for the football squad, yet it puzzled and annoyed me that he was haughty and seemed unnecessarily full of himself. Sure, he had plenty to be proud about. His appearance reminded me of David Clayton-Thomas, the lead singer of Blood, Sweat & Tears, or a more handsome version of Babe Ruth. From what little I could observe, he appeared less than completely trustworthy, and as the youngest of accomplished and admired siblings, perhaps he had gotten away with many things during his upbringing. Yet his looks and brash personality attracted plenty of attention from the array of lovely females at Central. One of the Tiger football players noted that Ed dated so many different beauties that members of the team would wonder which one would be his next "flavor of the month." Perhaps he was our school's Yul Brynner from the musical *The King and I*, simply needing to constantly flit from one lady flower to another, enjoying each visit but quickly departing to sip affection from another tasty bloom.

While Ed buzzed around the ladies, the third year at Central brought some pleasant changes for me, a boy who, now, wasn't named Sue. Under mathematics teacher Oma McNabb, I found unexpected, eureka success in her tough Geometry class. My teacher for English that year, Billie Chrisman, also helped turn on a lightbulb in my wandering junior mind. She required all students to explain a book of their choice to the entire class. Simply due to its strange title and unique cover art, I selected the 1961 nonfiction book *Black Like Me* by journalist John Howard Griffin. Its subject was racial injustice in the Deep South, and it struck home, delivering a lasting understanding about the reality of oppression. Though Central was integrated, the two races rarely mingled. To his benefit, Ed was always around the many talented black athletes, and he likely developed authentic friendships with them. But except for freshman band and one art class, I never had African Americans in my classes or as close friends, and I said hello to only a few during the hourly scamper to the next room of learning. It was my loss to never see the world of Central from the perspective of a Rutherford County student who wasn't white.

While knowing of the confusing and sad reality separating the races at Central, I discovered the joys of creating art. With the encouragement from the school's superb Art teacher, Jean Craig, I was inspired to imitate the psychedelic work of Peter Max and the exuberant posters advertising concerts held at the Fillmore Auditorium, rock music's sacred temple in the hippie, counter-culture scene of Haight-Asbury in San Francisco. Following so many unspoken rules, macho jocks like Ed would never consider enrolling in this "artistic" class, though it was their loss as creativity was flamed in this course like the most competitive of any Olympic sport. Mrs. Craig was delighted when I prepared flamboyant advertisement posters for dances sponsored by my fraternity. She introduced everyone in our class to many beautiful types of art, notably a sculptural form which could enclose, protect, and if exceptional, inspire. It was called architecture. Little did I know that this artistic realm would become my future.

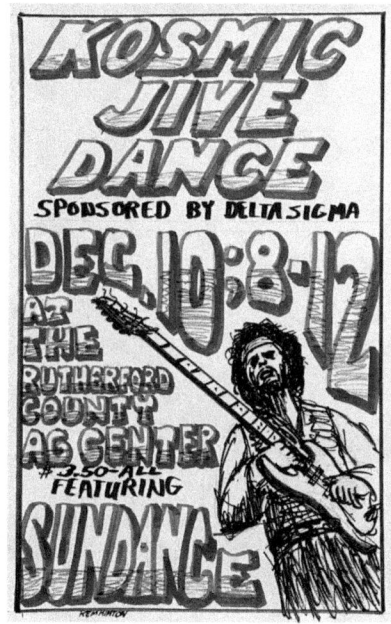
Dance poster

Perhaps the most fascinating course in my junior year was Chemistry. I vaguely remember Ed in that classroom, though I'm not completely sure. If he was, he must have been on good behavior due to the teacher's total command of the room. Florence Sublett had a stern yet pleasant personality and instilled in many of us a permanent curiosity and sense of wonder about chemical compounds, formulas, and processes. My lab partner was a cute, tall brunette who lived on a farm near the outer edge of Rutherford County. She teased my unusual name, calling me "Kemistry," yet I enjoyed her delightful personality as we tried to avoid blowing up the place while performing numerous, sometimes dangerous experiments.

CHAPTER 4
Senior Year

The fourth, final year at Central was transformative, with personal and educational encounters that would guide members of our senior class into a bright or not so bright future. For me and countless others, this year would be the best. I assume the same was true for Ed as his work with the football team helped the Central Tigers prepare for the school's final season, and his skills were noticed by several universities needing the finest of gridiron trainers. With encouragement of the best friend of my chemistry lab partner that previous year, I asked the country gal for a date to attend a James Gang concert in MTSU's small, concrete block-walled, acoustically horrible gymnasium. During that first date on a school night, the two of us learned how rock concerts in the wrong setting could be absolutely ecstatic while also delivering temporary deafness. The following day in classes, both of us heard nothing and saw only lips moving.

All classes in my Year Four seemed enjoyable and without pressure to ace as my decent grades already guaranteed graduation in the following May. One class that qualified as particularly memorable was Senior 401 English, and here I sought an A. Ed and I were both in this "homeroom" first session of the school day, along with the attractive brunette I continued to ask out. There were many excellent teachers at this public school, and the homeroom instructor, Eloise Womack, was one of the very best, displaying a matron-like, dignified personality and sincere interest in her pupils.

To encourage the reading of Chaucer and others of his era, Mrs. Womack organized an annual "Medieval Banquet" at the local agricultural center, requiring students to show up as a character from the Middle Ages. It was one of the highlights of fall 1971, with girls dressed as innocent maidens, nuns, and witches, and boys dressed as peasants, knights, and royalty. I came as an archer from Robin Hood's band of merry men, and Ed appeared as a hooded executioner, the dreaded one who chopped off heads.

Banquet with Ed in hood

Even during that hilarious event and other class interactions, Ed and I rarely conversed. We had nothing in common, or so it would seem. He had a few dates with a sweet redhead I had known for many years. As a compliment to his work with the football team and because of her popularity and valuable contributions to our senior class, she was included with other attendants on the parade float with the celebrated homecoming queen, all of these lovely ones dressed in flowing gowns. As class vice-president, I rode on our senior float nearby. Except for Ed's looks, I wondered what on Earth this girl saw in him. With the feminine crowd, good-looking Ed was perhaps charming, witty, and desirable, and maybe there was an allure to stepping out with this wealthy "bad boy." However, one of my senior year teachers

Homecoming queen and attendants   *Seniors homecoming float*

commented that well-dressed, smooth-talkin' Ed was Central's version of Eddie Haskell, the two-faced character in the syndicated television series, *Leave It to Beaver*.

My relationship with the brunette became a serious one, an unusual pairing of a suburban boy courting a rural gal. I adored her family, and we became very close, "going steady." Both of us joined a small group of managers who assisted the school's award-winning Central Tiger marching band. While Ed was working to help the football team during halftime, this juggernaut of 100 musicians entertained the large crowds who attended the Friday night contests. The female band managers handled scores and instructions of the complex routines, while the guys were tasked with carrying the "plume box," a large wood container full of feathery accents worn on the top of every marcher's distinctive band hat. More exciting for us was to handle and place warming cloaks on the exposed shoulders of the often-chilled yet adorable, barely-clad majorettes.

In February 1972, my steady girl's sorority had its annual wintertime formal dance. For this special event, she selected a light purple dress. To honor her with the mandatory corsage, I was obligated to purchase something that would match her distinctive attire. Though not my favorite hue, purple combines the energy of red with the calm of blue, and it represents luxury, prominence, and to some, gratitude. In ancient days, Tyrian purple dye was extremely expensive, and clothing toned with its distinctive color was reserved for royalty and those of very high status. For this contemporary occasion, the local florist composed a spray of lavender orchids. My investment in this expensive corsage was worth it, for once I pinned it on my date in front of her mom and dad, its distinctive color complemented by the shimmer of her amethyst earrings and the alluring scent of a heavenly perfume made her absolutely enchanting. I stepped back to savor the beauty of the moment.

Many years later, the color and name of the luxurious purple corsage would reappear, yet these would not be my girlfriend's. Instead, financial blooms would be made by the eventually murdered Ed and his compatriots. They would invent, cultivate, hybridize, and eventually sell a bouquet of insurance products that would have a different aroma, one of cold, hard cash.

Though Ed and I shared an after-school service activity that final year in the Key Club, we almost never spoke. However, we separately drove to a destination of a most pleasant kind. On the weekends, I would borrow my parents' 1965 turquoise Chevrolet four-door Impala to fetch my girl and enjoy dinner and a movie or a dance. Before returning her home, I would often take a special detour, one to a flat wonderland of romance. With his date, whoever it was that evening, Ed would also take that detour.

Ed and Key Club sweetheart

NETHER LAND

Only two blocks east from Central, the MTSU campus possessed at the far northeast corner of its sprawling property an enormous parking lot for commuting students. Normally full during daytime, it was completely deserted and pitch black at night. Because this large automotive amenity was on university property, the local police never came by. Campus police had other things to do or really didn't care. Thus, this spacious paved parcel was a neutral zone, the ideal location for maturing teenage couples to park, kiss, and perhaps do more.

Many of my fellow students would also take the detour to this popular, safe, and convenient spot, especially on the weekends. Vehicles would settle at different positions across the vast asphalt pasture, politely distanced from one another for privacy. The majority of cars making the pilgrimage had four doors with bench rows in the front and back, an arrangement preferred by eager boys for their dating excursions when compared to the hard-to-

Asphalt Heaven, bottom right

accomplish-anything bucket seats. All had radios, and a few of us provided romantic music for our sweethearts from still-new 8-track tape players. Although a fan of psychedelic and hard rock, I compromised and provided my girl's preferred tunes by "Top Ten" groups including Chicago, Three Dog Night, James Taylor, and the Fifth Dimension. Like all other young males, we would do whatever it took to provide an appropriate mood in which to gain affection. As one glanced across the array of silent vehicles on cold winter nights, steamed windows and absence of silhouettes confirmed cozy romantic activities within. Familiar with the parked vehicles of fellow students, I recall seeing Ed's silent and darkened vehicle in the distance, never knowing which favored flavor was with him that evening. The top on his distinctive, blue Pontiac Bonneville convertible was up, but the occupants were not.

Muntz 8-track

When I returned my girl to her home in the country, I would make the long drive back to mine, enjoying ear-damaging sonic delights by Jimi Hendrix, The Who, and Led Zeppelin. Years later in 1976, singer Bob Seger hoped in his hit "Night Moves" that everyone in that safe parking lot had gotten their share; one year later, a vocal duet by Meatloaf and Ellen Foley more accurately described the experience of some lads in the undisturbed, dimly-lit setting, perhaps seeing *Paradise by the Dashboard Light.*

Another time returning from my girl's place on a pleasant Sunday afternoon, I encountered a large herd of Jersey cows on both sides of the country road. They had escaped a farmer's fenced property.

Stupid me didn't slow down enough, honked the horn, and was shocked when one of the startled beasts jumped onto the middle of my path. I turned sharply to avoid the collision, but the swerved back side of my vehicle slammed into one of the frightened cows. When I tried to overcorrect, I ended up in the opposite ditch. Fortunately, I was still moving at a decent speed, gunned the engine, and made it back upward and onto the pavement. Looking back in the rear-view mirror, I noticed that no cow was down on the road, and therefore the one I hit must have been okay. When I reached home, my dad scrutinized the damaged car. With an enormous dent along the side, it was not okay, and given the look on his face, I was also not going to be okay. It was sour payment for a daytime trip just to see the brunette. Yet at the time, it was worth it.

Poster, 1972

One morning as graduation approached, Ed passed my girlfriend and me in the long school hallway, confidently holding up two fingers with a smile, or more precisely a smirk. It seemed an odd display of the ubiquitous "peace sign" by this jock. My sweetheart soon admitted that Ed had been calling her, and his hand signal was a reminder that she had promised to step out with him within two weeks. I was furious and pondered retaliation, but my girl assured me that her affection for me was solid and that I had nothing to worry about. Ed was a big guy, and taking him on would probably be a very big mistake. I was most grateful for her honest explanation about the hallway exchange and also for her

SENIOR YEAR

(momentary) faithfulness. Yet his hallway twin-finger seductive message to her was burned like a cattle brand on my mind, searing my ego. It would remain there, simmering as a permanent grudge.

A week later, the same thing happened, and yet when Ed passed us this time with the familiar hand gesture, I smiled and returned the same and said, "Peace." He didn't appreciate it one bit, and I expected an immediate physical confrontation. But she had turned him down, and he had plenty of other young ladies to chase and would just fly to other flowers for tasty satisfaction.

When the last edition of the school newspaper *The Hi-Lights* was published on May 19 that year, it included "last wills and testaments," short and silly statements from each senior ready to move on with life. Each bestowed some item or quality to another student, usually a younger one. The issue also featured hilarious predictions by the paper's female editors on what each graduate would do in life. Given artistic awards I had received, this group felt I would eventually repaint the ceiling of the Sistine Chapel, and due to his playboy reputation and appreciation of all female flavors, Ed would become the master of ceremonies for the Miss Universe beauty pageant. Although we never discussed it, or anything else, I believe both of us were very pleased with these imagined futures.

Art awards: author and talented friend Billy

At graduation on May 26, my girl and I were still tight, having

NETHER LAND

Ed in 1972 yearbook

enjoyed the last lessons in our assorted classrooms before that pivotal moment when diplomas were presented and, like it or not, everything would change. Our graduating class was so big that the event, like all of Central's football games, was held at the MTSU football stadium, and though no one really listened to the speeches on that pleasant afternoon, the occasion was thrilling as each person paraded across the stage to receive their ticket from principal John Swafford. A few made obnoxious gestures, knowing there could now be no penalty. My "main squeeze" and I embraced in our black gowns and shared warm kisses for the future. We were two of the nearly 500 that day, the largest high school graduation in Tennessee history.

The important senior prom later that night was a sad flop. With bad advice from some, commanding teacher and class advisor Sam Moore selected an "uncool" seated band of the Lawrence Welk genre for the dance. Most students wanted a contemporary group that would play the loud, popular rock 'n' roll music of the day, the hip sounds that constantly flowed from our radios, record turntables, and 8-track players. Only a few days before graduation, big bear Moore directly confronted me one morning, angrily accusing me of writing an anonymous letter that had just been

POSTSCRIPT 1972

1972 CHS yearbook

SENIOR YEAR

published in the school paper. I had not written it, but I agreed with its criticism of the selected musical group for the final event of our high school experience. On that prom night after posing for the obligatory dance portrait, most disappointed students therefore left to attend off-campus parties. Some headed to the nearby, quiet asphalt heaven. Fortunately, the photo of each sharply-dressed couple possessed a message at the bottom, one declaring when the 10-year reunion of our class would occur. It was wisely scheduled for a future time when most grads would likely be back in town to see their parents and others during the wintertime events of Christmas and New Year's Eve. Overall, graduation had been exhilarating, and most attending the prom would keep that photo and anticipate a party so many years away. The coming decade would pass much more quickly than any of us imagined. We would always call the cherished Southern town in which we had matured as the "Boro."

Prom message

Most graduates of Central would continue education at a college or university near or far away. Ed received a valuable full scholarship to be a football trainer at the University of Tennessee, and the star player Hollywood received the same to play on the awesome Big Orange team. I had mediocre grades, so there was no financial support, yet I also planned to head to Knoxville and enroll in the five-year Bachelor of Architecture program. There were several at Central who sadly never found their way, but most of us taught at that gigantic place would lead meaningful lives and achieve success. Our class included those who would go on to careers that included attorney, lobbyist, communications manager, travel agent, toy creator, horticulturalist, anesthesiologist, assisted-living facilities executive, police officer, funeral director, nurse, school superintendent, teacher, bank branch president, noted collector, fashion buyer, counselor, college professor, architect, printing manager, football coach, surgeon, contractor, sailor, engineer, farmer, writer, insurance agent, retail store owner, landlord, stock broker, dental issues and equipment "fixer," award-winning television and film graphics pioneer, PhD art curator and author, nationally-known stage manager (for the Emmy, Grammy, Oscar, and Tony award shows), caterer, banker, landscaper, preacher, and inventive 50-caliber semi-automatic rifle maker.

Central High class of 1972

The next year, Central High was closed because it could not accommodate Rutherford County's growing student population. The once single, powerful collection of capabilities in all realms—especially sports—was chopped in half to form two new schools named Oakland and Riverdale. Loyalty to one school was eradicated and sadly but inevitably replaced with a rivalry between two brand new places. The educational facility in which we had matured in

SENIOR YEAR

grades 9-12 became the county's junior high and renamed Central Middle. Although students in the ninth, tenth, and eleventh grades were probably excited to start the heritage of one of the replacement high schools, to me and others it seemed incomplete when compared to the unified journey we experienced. How fortunate our graduating class had been to share at the once-only secondary school in Rutherford County the adventures, lessons, shenanigans, maturation, romances, rivalries, and relationships in a corridor which seemed to never end. School newspaper editors Beth and Carolyn had already delivered a somber farewell entitled "The End," and as we closed each of our yearbooks filled with pages of sincere, touching, or goofy messages for the future from fellow students, that important chapter in our lives ended. One year earlier, pop singer Carly Simon had predicted at the end of her lovely song "Anticipation" that "these are the good old days." Looking back, she was correct.

That summer, I worked at the preeminent photography portrait studio in the Boro. As a high school graduate with few skills regarding that specialized trade, I was happy to perform any task. The entrepreneur owner, Ed Delbridge, was one of the most prosperous men in town, and from his encouraging staff I leaned a great deal about photography, print making, and portrait finishing. He was a class act, and his grown daughters were delightful, smart, kind, and stunningly attractive. One of them, Sally, had been the homecoming queen at Central in the fall of 1970 when my friend Hollywood led the Tigers to become state champions. She had just

The End

Last rites for Central High

finished her freshman year at UT and was a "little sister" of one of the many campus fraternities. She casually mentioned the young men in that group and suggested I should consider them if I wanted to join a Greek society. I was extremely flattered because she was one of the most charming girls in all of Rutherford County. This was reinforced by a close friend of my steady love's older sister, as she was dating a member of the same chapter and encouraged me to consider his group. As my journey at UT was about to begin, her fella, also named Ed, invited me to visit the house during fall rush. Given his sincerity and the fraternity's choice of Sally, I knew that was where I wanted to be. (Later, to my great fortune, he would become my designated "big brother.")

CHAPTER 5
First College Days

In the fall of 1972, Ed, Hollywood, and I along with several other Central graduates, headed to Knoxville to enroll at the University of Tennessee. The two athletes moved into Gibbs Hall, the elite dorm full of jocks known for its generous, all-one-could-eat, athlete-nourishing cafeteria. I lived first in the old, non-air-conditioned Greve Hall, later relocating to the newer Carrick Hall to room with another Central graduate, Norman. Fraternity rush occurred

University of Tennessee UT

during the first weeks of school, and I headed directly to the chapter house of my first choice. I was concerned that Ed might attempt to join the same organization, yet fortunately he was interested in a different one. Both of us pledged and soon became fraternity boys at competing groups on the university's Greek Row.

Though now three hours away by car, I stayed in close contact with the girl back home. She enrolled at MTSU, and our long-distance relationship initially seemed solid. I made several trips that first quarter from Knoxville back to the Boro to enjoy home cooking, be spoiled by my mom, obtain car keys from my dad, and most importantly, go see my gal. Other than those visits, I attempted to strengthen our relationship with love letters, floral bouquets, and (back then) expensive weekend phone calls. Hearing each other's longing voices kept the romance hot, or so it seemed.

The Big Orange football games I attended that first year were fabulous and unforgettable. The most spectacular Saturday event, it consumed everyone and everything on the university campus. At that time, the giant horseshoe-shaped stadium had north end zone bleachers and two cantilevered upper decks that seemed to defy gravity, and students would fill both levels in reserved sections on the east side. Unlike the rambunctious and vocal students, alumni and season ticket holders filled the west, afternoon-shaded part of Neyland Stadium, remaining polite except in reaction to awful calls by referees. Beloved "Voice of the Vols" radio announcer John Ward delighted fans tuned to the radio with his distinctive "Give him six. Touchdown!" announcement. Another message during the games would be to hear the announcer clearly thank "a host of Volunteers" when the players pounced en masse onto the opposing team. There was also another host of volunteers, this time students, who utilized the liquids of Jack Daniels whiskey or Bacardi rum in smuggled flasks to mix with Coca Cola or Sprite and thus provide soothing libations for a most enjoyable stadium experience. Fraternity members like me were expected to dress properly for the occasion, and many of their dates attended the early, warm fall games in that very appealing of feminine attires, the "sun dress."

U.T. football managers (Ed back row far right)   UT

Using binoculars, I watched Hollywood on the offensive team, proud to see this hometown boy from the Boro on the still-controversial Tarton Turf playing field. I also saw Ed on the sidelines assisting players throughout the game. Two important moments occurred at the stadium that fall season. It was the first time a game was played at night under the newly-installed flood lights, and it would also witness the premier by the university's huge Pride of the Southland Band of a 1967 tune composed by Felice and Boudleaux Bryant. The couple's "Rocky Top" soon became the celebratory "anthem" of the university. Later, it would also become one of the Volunteer State's official songs.

The University of Tennessee campus was so large that it had a

NETHER LAND

Ayers Hall UT

bus system, one that took students living in Hess Hall, known as "The Zoo," and those residing in dorms Carrick, Humes, Reese, Morrill, and Andy Holt Apartments to the distant academic buildings that crowned the historic center of the university known as "The Hill." This summit plateau featured the institution's most prominent landmark structure, Ayers Hall. On the way, the buses passed Fraternity Row and the athletic dorm, and I recall only a few times when Ed and I would be on the same vehicle making the same long journey. During that first fall quarter, I once saw him wearing "clog" shoes, the latest trend in men's footwear, consisting of a big, audible wooden heel and a wide leather top strap. He was again establishing his well-dressed reputation, and this slipper was stylish. I never purchased a pair and thought those clogs looked like clunky and oversized Dutch flip-flops.

That first year, near tragedy struck. On a cold Sunday evening in November, my girlfriend cried during our weekly phone call, confessing that she had stepped out with Ed that weekend. He had been in the Boro while I had stayed in Knoxville to prepare for an important trigonometry test. As singer Gordon Lightfoot would say, that pesky Ed had again "bin creepin" and succeeded in luring her time and affection. She said she was sorry and didn't want to lose me. I was blindsided, especially given her firm denial of his temptation only a few months earlier. In frustration, I asked, "Why?" Ever the playboy, Ed was continuing to buzz around from one flower to another, and he'd landed on one I thought was unavailable. I should have hung up the phone but was naïve and forgiving, imagining a future with her and wanting to continue our special times alone. Her apology seemed sincere, though it flowed from 180 miles away along very thin and fragile copper wires. I stayed on the phone and with her. The distance-challenged relationship seemed to strengthen, and we jointly expressed a sincere commitment to each other. I later wondered if Ed had worn those unusual clogs on the now-exposed date in the Boro with my lady.

In his freshman year, it is probable that Ed immediately took business courses. In my freshman curriculum, there was only one class related to architecture, and it was a mediocre introduction by an upperclassman to a career I wasn't yet convinced I should pursue. My other courses included calculus, English, art history, studio art, and an extremely difficult engineering "statics" class taught by Raymon Shobe, a serious older instructor who was in many ways the equivalent of the crusty Professor Kingsfield, played by memorable actor John Houseman, who ruthlessly taunted first year law school students in the movie *The Paper Chase*. I remembered an embarrassing example when, after several of us had given an incorrect answer on a test about tension and compression, Professor Shobe asked us to stand. He slowly placed in each of our hands a short piece of twine and then asked us to demonstrate how one would push on this string. It was humiliating and totally effective. In his class, we were also required to solve complicated structural problems using a slide rule; calculators made by Hewlett-Packard and Texas Instruments were still new, very expensive, and absolutely prohibited in his classroom.

The bus stop near my dorm of Carrick Hall was always crowded with impatient students waiting for the next vehicle. One rainy, miserable morning in February, I was there, quite grumpy to be headed to a dreaded, way-too-early 7:50 am calculus class in Ayers when I saw another individual in our assembly of half-asleep zombies. She was a stunning blonde, as beautiful as any I'd ever seen. We both stepped on the bus, and I wondered who she was. She didn't notice me, and I had no reason to say hello or make any overture. At that time, I wasn't interested in other girls, but if I thought I had a chance with this special one, I might have adjusted the knob on my fidelity meter several notches lower. I didn't know it, but fortunately that dial of faithfulness would soon be turned off—not by me but for me. Years later, I would have the chance to ask this mysterious girl for a date, changing my life in most wonderful ways.

Academic successes in that initial college year were crucial, as the final grade point would determine if I would be admitted to the professional architecture program that officially began in the sophomore year. Of the one hundred freshmen interested in the career, only fifty of us would make the cut. I have no idea which freshman courses Ed took, but I am quite sure he was in the Economics 101 class, a requirement in his curriculum taught by the university's most flamboyant, entertaining professor, Dr. Tony

FIRST COLLEGE DAYS

Spiva. I took that class as an elective, quickly learning that "bizzness" was definitely not for me.

As our freshman year ended, Ed and I both headed home for the summer. Before leaving, I was invited to a freshman scholastic fraternity. At the honorary dinner in the University Center, I was seated next to a tall and imposing man with orange hair. This university official delighted all of us at the eight-person table with enthusiastic interest, charm, and an authentic curiosity about our separate journeys and individual successes in the first year away from home. How fortunate I was to sit beside and soak up an hour of wisdom from this legend who had formerly served as the university president. During his tenure, enrollment increased threefold, faculty and staff doubled, and government funding increased fourfold as he worked tirelessly to make the university Tennessee's finest. His sincere, personal encouragement for me to pursue and excel in the architecture curriculum was unforgettable. His name was Andy Holt.

Blackberries　　　　　UNS

Summertime found me in the Boro continuing with the country girl while working again at the portrait studio. She never mentioned Ed, and we had a wonderful three-month break from the books. My squeeze taught me how to make "balloon wine" from an unending supply of blackberries on her parents' farm. I had no idea that these juicy treats bursting of sweet and tart flavor were packed with vitamins, rich in fiber, and full of all kinds of good ingredients including the anthocyanin pigments that delivered the inky, deep purple color. All I cared about was that the luscious berries were delicious, almost as good as kisses we exchanged while harvesting the purple nuggets.

After our solution of crushed berries, water, and yeast began to ferment in a ceramic crockpot, we poured the bubbling liquid into a clear glass jug and covered the narrow opening with a large balloon. We then witnessed the fascinating process as the rubber item would, again and again, slowly expand then rapidly release carbon dioxide from the alcoholic process. After a few weeks, the homemade beverage was ready to consume, and we'd bottle the sweet liquid for enjoyment at a later time, perhaps in a darkened parking lot. Making both blackberry and grape wine was simple, and I would always owe the farm gal with tanned, PB legs who had been my lab partner for showing me this process, one that would have surely delighted and perhaps intoxicated our high school chemistry teacher. (Later in life, I would again make wine using juices from Italy, France, and Spain, and multiple bottles of these better vintages would silently rest in a cellar beside a few saved from that flavorful summer more than fifty years earlier.)

When the exquisite summer activities ended and August of 1973 arrived, it was time for me to travel back to Knoxville. I saw Ed in a car hauling his stuff back to campus as his vehicle passed my dad and me on east-bound I-40. Separately, Ed and I began our second year. Mine would be quite intense as I was totally focused on coursework, receiving a first full taste of architectural design in a makeshift studio occupying a tall, former basketball practice area in the university's old Alumni Gym. At the first few football games, I again noticed Ed's work on the field. He must have enjoyed being on the artificial turf surrounded by thousands of noisy, appreciative fans as the Vols were undefeated as the second half of the football season began.

FIRST COLLEGE DAYS

When the middle of October arrived, the president of my fraternity urged me to join him and attend an evening event, one of the university's frequent "Issues" programs, where lively and often controversial speakers would be invited to the campus to share their experiences and insight. That Monday night on October 15, 1973, the topic was "Who Killed JFK?" a sobering presentation by journalist Robert Katz. I remember well the moment ten years earlier on November 22, 1963, when fellow classmate Janet Marcum, one of the few Catholic students at Campus School, came running into our fourth-grade classroom and sobbed as she told us the breaking news from Dallas. Perhaps the saddest event in the lives of our then-innocent generation, the harsh reality of the murder of our nation's leader emotionally scarred all of us.

Young investigator Katz was challenging the findings in the Warren Report on the assassination of President Kennedy. To the stunned audience in a packed room at the University Center, he showed a bootleg version of the famous Zapruder film, the only known movie documenting the chilling murder. Most of us had seen the series of single images, but now we were watching a moving, completely soundless documentary of what happened in vivid, terrifying color. He then reran it slowly, and with the horror of the President's head blasted off and his body's terrible smacking backwards, he convinced the shocked and totally silent room of the likely truth that another gunman, perhaps on the "Grassy Knoll," must have joined Lee Harvey Oswald in the dastardly deed. (Katz's surreal presentation was likely not attended by Ed or anyone else involved with the football team, as the undefeated Vols were facing in just five days our most-hated—at the time also undefeated—rival, Alabama. After a disastrous fourth quarter for the Vols that

Saturday, the Crimson Tide won, 42 to 21.)

One month after witnessing the disturbing Zapruder movie, the solemn tenth anniversary of President Kennedy's assassination was observed by his lifelong rival, President Richard Nixon. This other leader had overseen the remarkable Moon landing on July 20, 1969, and yet due to his unwarranted fear of not being reelected, the stupidity of Watergate break-ins, and other outrageous acts, Nixon would later fully disgrace the presidential office, politically and morally crash on Earth, and resign. How disgusting it was that the commemorative metal plaque on the remaining half of the Eagle spacecraft, left as evidence when astronauts Armstrong and Aldrin blasted off, would forever mark the landing site of one of mankind's most stupendous achievements, and yet it would have Nixon's dishonorable name on it instead of the president who had challenged a nation to reach for the stars. For a moment, I wished I could journey to that enshrined spot on the Moon, find the silent vehicle and its important plaque, and with the same ballpoint pen that had changed the spelling of the name on my birth certificate, quickly scratch across one president's name and, for eternity, inscribe the one who deserved to be honored, John F. Kennedy.

I had reasonable success in classes during that fall quarter at UT. Other than attending football and basketball games, fraternity beer busts, and rerun movies shown at the Campus Center, I was focused entirely on the difficult math, engineering, structural, and mechanical systems courses in the grueling architectural curriculum. Similar to many of my age at UT, I did not possess a car. Walking all over the enormous campus made for great leg muscles, but it stifled any opportunity to explore Knoxville or do other activities requiring four wheels instead of two legs. I continued to hitch rides

FIRST COLLEGE DAYS

back to the Boro for time with my steady girl, and she made the same three-hour trip to enjoy a few weekend visits with me. Things were quite nice.

A month before 1974 arrived, the world seemed to shake, foreshadowing other tremors in politics, student misbehavior, and my personal situation. In the wee morning hours of Friday, November 30, 1973, I was sitting in bed in Morrill Hall composing a letter to Debi, a very close friend back in the Boro when a thundering earthquake hit the Knoxville community at 1:48 a.m. It was composed of two jolts of rumbles and terrifying shaking, setting off every fire alarm on campus and waking students in all the dorms. Disoriented, they hurried down exit stairs and onto surrounding areas. The same thumping happened across campus where Ed and every jock were also rudely awakened. Because there were no female dorms on his side of the university, we boys in the common-folk coed dorms were delighted to see all the girls in quickly-donned bedroom robes or jammies without makeup or normally nice hairdos. Our large assembly of impatient ones uncomfortably waited until the coast was clear, and then, still-dazed and half-asleep, shuffled back inside. Fortunately, only minor damage was reported later that day from this Richter scale 4.7 shake-up of the Big Orange.

When 1973 became 1974, things started out quietly, then instantly erupted. College students have always been copycats, imitating the actions of others at faraway places of intellectual freedom. How fun it was to learn of the silly "panty raids" of the 1950s and then compare them to the more recent, serious Vietnam War protests. It was no complete surprise that another craze would sprout, this one beginning above the Mason-Dixon Line in 1972 and 1973 at Notre Dame University when a single male or a small group of them ran quickly across the school grounds completely naked. The outrageous prank became known as "streaking" due to the high-speed sprinting of one who wanted to share the glory of his birthday clothes but hopefully not share any recognizable facial features. Most onlookers peered at the bottom half of the rapidly-moving individual, unless it was a rare female, and then glances would be more bodily comprehensive. Streaking quickly became the latest fad at other institutions, and by the end of 1973, *Time* magazine firmly acknowledged this growing national silliness at American's fine institutions of higher learning.

In January 1974, the national press called streaking a "national epidemic," and the well-informed student body in Knoxville was certainly not going to let everyone else have all the fun. A few streakers had run through the large campus, and most felt this was just a frustrated one letting off steam... and clothes. However, on the early evening of Thursday, February 28, 1974, the obnoxious, countrywide fad finally hit the Tennessee campus like an explosion. Informed and encouraged by radio broadcasters, a tidal wave of students swept out of the dorms and hurried to the wild festivities and already swelling sea of exuberant, joyful students on Cumberland Avenue, the nearby zone of student-targeted stores, inexpensive restaurants, and legendary beer joints. Like so many, I foolishly entered the mass of excited students in the completely overwhelmed and overcrowded street. Maneuvering to the center of the mass, I glanced upward to the south and saw a large group of my fraternity members on the roof of a closed gas station. They had hurried to grab good seats and were having an evening like no other, peering down from the excellent observation setting to cheer at the frequent male streaker or a single-file group of naked bubbas as they hurriedly moved through the ever-expanding noisy and delighted crowd. Turning to the north, I saw on adjacent rooftops many guys and one girl disrobe completely. Having learned photography the summer before, I captured with my trusty Minolta 35mm camera assorted images of naked ones, although the precise term would have been comedian Lewis Grizzard's more accurate definition of one without clothes and up to no good as "nekkid." It was initially hilarious, yet caught in the sea of people, I became lodged between others and was unable to control my direction, and I was literally and physically moved with the crowd. It was not a pleasant experience, like being in a slowly-ebbing stampede.

Mob of excited students

I could not confirm that Ed and the athletes from Gibbs Hall had sprinted across the campus to join the noisy assembly whose decibel level rivaled the excitement and cheering during a football contest between the Vols and Alabama. But they surely must have arrived to watch or join in, as almost every young person in the surrounding zip code had. Nothing like this had ever occurred in

NETHER LAND

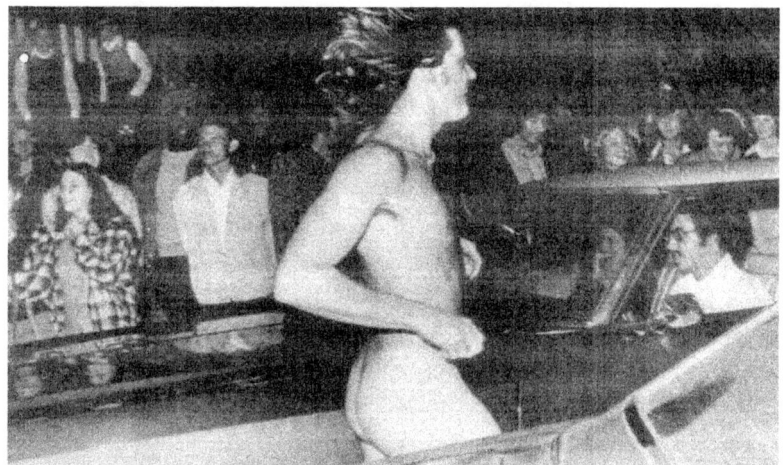

Streaker in Daily Beacon   UT

Knoxville. Streakers continued to delight the cheering spectators long after midnight.

Well into the morning hours, the last tired and intoxicated revelers finally dispersed, and on the following days, many kids begged forgiveness from professors who faced empty classrooms. Fearing arrest for the nudity I captured on Ektachrome slide film, I waited several months before having the roll developed, and it indeed showed some rather juicy pictorial evidence from that crazy night. How amusing that Cumberland Avenue, the scene of the scandalous, rabblerousing event, had years earlier already been tagged with a most appropriate, now permanently validated nickname, "The Strip."

In the early spring of our sophomore year, I was back home

Cumberland Avenue

for the weekend and headed to my girl's dormitory. When I arrived at the pick-up area, she quickly entered the car, quietly positioned herself on the front bench seat, and turned to face me. I was expecting a moist kiss, but there was nothing encouraging on her look. I could tell that something was wrong. Although it wasn't funny at the time, I felt like Astro, the hilarious talking dog on the 1962 futuristic cartoon show *The Jetsons*, who met arriving doom by uttering, "Ruh-roh."

The Jetsons dog, Astro   IMA

She quietly said, "I want to break up." Although taken aback, I didn't ask why. It was clear that affection for me had vanished. There was nothing to say. Silence in the car confirmed that the love and fun was now over. It was honorable that she ended the relationship in person. After a few moments of dead quiet, she climbed out of my car and quickly walked back to her dorm. I later contemplated calling her but never did. It would have been a waste of time now that the fidelity meter had been cranked down and turned completely off. Afterwards, I learned that she may have needed freedom to accept a new suitor's invitation to his fraternity's "Old South Ball." If so, I had been ditched for a dance. Yet, who could blame her, as this was a celebrated springtime event, one where young college boys would don the officer uniforms of the

Pageantry of Old South Ball

Confederacy and young women in flowing, delicate dresses would parade as glorious Southern belles. Perhaps the guy would be lucky and my ex would wear a light purple gown to the annual, make-believe event. Given her new availability, I also assumed that she and Ed would eventually reconnect.

Couple at Old South gala

I experienced mixed memories, wondering if that deep cut of rejection would heal as a ribbon of past affection or a scar of betrayal. Regarding romance and life, we now sped down different highways, ending in distinct destinations where each needed to be. Eventually, the hometown girl would fully blossom with two impressive university degrees and become a wife, mother, and beloved school teacher.

The end of that most unusual sophomore year concluded with a fascinating, national news twist. Washington Post journalist Bob Woodward came to the campus to speak on April 18, 1974. A first excerpt of two parts of his blockbuster book *All the President's Men* had just been featured in *Playboy*, the monthly publication known for its scholarly articles, fashion tips, and photos of nature in all her glory. That particular issue of the magazine was enjoyed by me and—I am reasonably sure—by Ed and every athlete for its cultural and artistic values and for this highly anticipated expose on the disgusting saga of Watergate. Woodward's revelations that Thursday night were shocking and equally hilarious, exposing the truth about President Nixon, Deep Throat, and the buffoons involved in the Watergate scandal. I had just joined the "Issues Committee," an undergraduate

FIRST COLLEGE DAYS

Bob Woodward with students UT

group charged with suggesting and hosting prominent speakers of interest like Woodward, especially ones that might be similarly controversial. After Woodward's presentation, our small team of hospitality escorted him back to his hotel, the nearby Sheraton Campus Inn. The university official in charge of the committee and I finally departed late in the evening, leaving handsome, single Woodward pleasantly stuck with several attractive coeds in the hotel's lounge.

Knowing I had some graphic capabilities, the campus representative suggested a deal for the coming year. She would arrange for me to receive art materials at the university's main bookstore, and for that I would prepare a publicity poster for each celebrity scheduled for the next academic year. Her flattery and trust in my skills were nice touches upon which to head westward to the Boro for summer break.

The hot months in the middle of 1974 found both Ed and me in the Boro, although I don't know if he had a job or just slummed at home. It had been no fun for me being tossed to the side, but the unilaterally wrecked romantic relationship had delivered an unexpected prize. I had been faithful, hadn't "fooled around," and had concentrated on the difficult architectural curriculum. The resulting towering grades from the first two years surprised—actually shocked—my parents. Having gained confidence and the valuable cushion of high marks, academic pressures eased, and my next three years in college were relaxed. I was free from the unavoidable pressure felt by so many to raise their grades.

Employment during this time away from academia found me working on a construction site at MTSU. Exposure to the process of making real structures from the illustrations on large pieces of paper called "blueprints" was a requirement for all students in UT's professional architecture program. It also ensured that bookworms like me who liked to draw pretty pictures gained a baptismal appreciation of the necessary complicated, messy, hot, loud, foul-smelling, dangerous, and brutal process in the "real world" to transform pencil and ink compositions prepared in nice, air-conditioned offices into actual buildings. My coworkers were on the rugged side, none with anything above a high school diploma. I was constantly singled out by these tough characters as the privileged "college boy." During lunch breaks, we occasionally bantered about politics, baseball teams, and the difference between the speed of a hot-rod Chevy Camaro and the maneuvering go-kart agility of a Porsche 911. These fellows preferred the drag racing rocket made in Detroit while I defended the sleek German missile. My time on the job and with these individuals again reminded me of the value—

frankly the urgency—of obtaining a university diploma. Regardless, working in construction and witnessing the building process firsthand taught me in only three months as much as two years in those classrooms in Knoxville.

Cruising around after work as a dateless wanderer in my hometown, I occasionally passed the spacious asphalt-paved field of dreams at night, but alone had no reason to enter. It was a quiet and lonely summer. I saw Ed one evening at the bar of the same country club where we had worked and played years ago. He had a lovely one sitting with him, the flavor of the summer or, perhaps, a steady lady. We did not speak. In forty years, Ed would be murdered.

CHAPTER 6
More at The Big Orange

The next academic year, Ed was housed again in Gibbs Hall, and I lived in my fraternity's chapter house on Volunteer Boulevard, its classic Doric columns and pediment not my favorite architectural style but certainly of an esteemed traditional appearance on Greek Row. Thirteen other chapter houses were also positioned in this zone of the big campus, each affiliated with a national fraternal organization. Most were filled by decent young men interested in the fellowship of others and housing in a smaller setting. But a few had knuckleheads determined to get into trouble at every opportunity, and these characters were notorious pains in the side of university officials. With a few exceptions, our house was always festive yet reasonably well-behaved. It was nothing like the side-splitting portrayal of fraternity life a few years later in the 1978 movie *National Lampoon's Animal House.*

One unsubstantiated rumor of a prank—at the time fun and immature yet now considered outlandish and even disgusting—told of an existing telescope in one of the chapter houses used by members to peer directly northward to a nearby female dorm, and that as self-trained astronomers, some were able to spot heavenly bodies in Knoxville's evening skies. Other crazy fads came and went, including the urgency among all fraternities, including mine and Ed's, to obtain the hard-to-get yet reportedly most wonderful liquid on the entire planet, Coors Beer from faraway Colorado. (Later in 1980, Lisa Birnbach would nail the traditions, mannerisms, preferences, and characteristics of fraternity boys in *The Official Preppy Handbook*, documenting the proper attire—starched khaki pants, more-heavily-starched cotton shirts, Sperry Top-Sider loafers, and either Ralph Lauren or Izod polo shirts—worn by most young men on Greek Row.)

It didn't help that Norman, my roommate then and for the previous two years, was extremely handsome, a heartthrob known by many of the 14,000-plus enrolled coeds. My notoriety from girls interested in this tall friend was "Oh, you are *his* roommate!" The two of us decided to journey to Auburn University and attend a pivotal Southeastern Conference football contest pitting the Vols against the AU Tigers. In the past, the Vols had always played the Tigers in Alabama at the large Legion Field stadium in Birmingham, yet this game would be the first time the matchup would be played on

UT/AU football program

Auburn's campus. Norman had an adorable date named Beth-Ethel, and she promised to secure a "cute one" for me. Unfortunately, a blind date can sometimes send a sudden wish for temporary loss of eyesight. For me, it would be an immediate, urgent prayer.

NETHER LAND

How fun it was on September 28, 1974, to sit in the 61,000-seat Jordan-Hare Stadium and witness Auburn repaying the embarrassing "Rain Game" defeat the year before in Knoxville, this time flipping the same numbers on the scoreboard, 21 to zip. The AU enrollment that fall was approximately 16,000, and the majority of them were attending this historic, first-time sporting event on their home turf. It appeared that half of them were attractive coeds, many wearing sun dresses. However, I was stuck with the promised date, tolerating the ballgame shellacking in that sold-out outdoor arena. Looks aren't everything, yet these Auburn girls seemed savvy, fun, smart, and quite attractive. Not so with my set-up companion. Puppies are cute, and I was/am no Robert Redford, yet as I glanced around, I wondered if the rumor (or joke) was true that, for diversity reasons and to deal with alumni parents, five surly, unpleasant, dumb, or homely women were admitted each year into the student body. My date was in that category. She was frumpy, uppity, and grumpy, and yet perhaps she felt equally betrayed by the matchmaker who had set up the mismatch pairing. Thus, it was a long, hot, and completely unpleasant afternoon for two youngsters who would have preferred to be with any other person and at any other place on the planet.

When the football game that afternoon was thankfully over, Ed and the downtrodden team loaded onto chartered buses and headed straight back to Knoxville. At least they escaped the rest of that unfortunate day. The Vols would later finish the season ranked 20th by the Associated Press and achieve a confirming win over the University of Maryland in the Liberty Bowl. But as for me that day, I was unfortunately glued to the seat in the elated Auburn stadium with Miss No Fun. My duty continued as I was obligated to treat this date to dinner and accompany her to a fraternity dance later that night. Albert Einstein would have been fascinated to witness time on that day, as defying the laws of physics, it seemed to painfully ooze at the speed of a disabled snail.

The return ride to "K-town" on the following day was quiet, and my roommate smiled the entire time, basking in the delightful time with his dreamy escort and joking with me about the twelve-hour misery that I had endured. An unpleasant Medusa in the land of otherwise delightful War Eagle girls had turned my highly anticipated weekend into stone. My only satisfaction was to tease back, reminding him that on the passenger's side of his red Nova, I was watching the asphalt highway whirl by only inches below a large hole in his vehicle's floorboard. The rusted opening served as the car's sole air conditioning on that scorching Sunday afternoon as we slowly made the agonizing 308-mile trip home from that damn "loveliest village on the plains."

After the brutal weekend at the start of my third year, the situation took a nice turn. As seasoned upperclassmen in completely different worlds, I had a comprehensive understanding of the university and its incredible variety of activities. Ed probably felt the same and additionally enjoyed all the phenomenal athletic facilities and participated in his fraternity's numerous intramural contests. I trust he also attended concerts in the acoustically miserable Stokely Athletic Center, including the second time that Elton John traveled to UT to provide, with the opening act by Kiki Dee, an astonishing performance on November 9, 1974. My pastime interests widened to cultural and political edges, and I especially enjoyed serving on and providing artwork for the Issues Committee.

With artistic supplies and artistic freedom, I designed publicity posters announcing upcoming visits by well-known speakers. These were printed and stapled on bulletin boards in dorms, dining halls, and classroom buildings. Yet the real treat was to be a taxi driver, serving as a chaperon for the famous, including safety advocate Ralph Nader, Watergate-stained John Dean, Future Shock author Alvin Toffler, Charles Manson's prosecutor Vincent Bugliosi, Israeli general Moshe Dyan, sex therapists Masters & Johnson,

Issues Committee poster

Black Panther leader Eldridge Cleaver, and Star Trek creator Gene Roddenberry. I asked most of them to sign the poster I had prepared, but grumpy Nader refused, saying he never, ever shared his autograph.

To widen the scope and attract others, especially the superb student athletes, our committee invited the legendary UCLA basketball coach John Wooden to share insights about his leadership techniques, athletes, and unmatched ten NCAA championships. In the distance, I saw Ed among all the jocks who had emptied Gibbs

Speaker for the athletes

Hall to join the packed crowd in Alumni Gym. The admired, wise sage Wooden delivered a memorable speech about the awesome power of teamwork and the lifelong value of individual integrity. "Reputation," the owlish coach asserted, "is what people think you are. Character is what you really are—something only you can really know."

Living in the fraternity house during my third year had been fabulous, and due to seniority, my roommate and I selected one of the best rooms for our fourth year. Daily meals provided by cooks Catherine and Jody were enjoyed in a large dining room, and during each meal, brothers continuously jabbed at each other, a practice called "giving grief." Guys received nicknames such as Casual Pierre, Bruder, 6-4, T.D., Galileo, Deda, Pud, Rella, Geech, Pablo, Roast Beef, Tater, Ink, Snake, Scott-Dawg, Red, Young Gnat, and some others not proper to disclose.

The real value of the fraternity was to be around and learn from the sharp "old heads," members a few years ahead who would later honor our organization by success in their chosen fields. Our place had some of the very best. One would become a United States senator, and another so financially successful that he would eventually purchase a professional football team. In addition, as predicted by my dad, most of my closest lifelong friends would come from the same-age "pledge brothers" who with me had joined the chapter as a freshman and remained to finish their coursework and receive a diploma. I suspect that the same was true with Ed.

During his time at UT, Ed seemed to have no problem finding new flavors to date on the enormous 910-acre campus. Occasionally Central graduates and others from the Boro on the sprawling Big Orange educational empire met and exchanged stories about the hometown and fellow high school friends. Several including me had noticed that Ed was now dating the head majorette of the university's marching band. While he was assisting the football team in the locker room, this slender beauty with golden hair would delight the gigantic crowd in Neyland Stadium as star of the band's halftime presentation. She rapidly twirled and launched her chrome baton almost into orbit, catching it every time with the grace of a ballerina. From a distance, it appeared that with this attractive and talented girl, Ed had secured the interest of the most skillful and flavorful. I'm quite sure he felt that beside her, he was one big man on the very big campus.

Formal dance, Ed bottom right

For me, finding a girl was a challenge. Between long hours at the architectural studio in Estabrook Hall, I attempted to invite a female companion for the multiple outings where one was expected to have a date: dances, movies, dinners, lectures, parties, concerts, and, of course, the unending supply of exciting athletic events. Fraternity beer busts and other encounters with young women in sororities improved my odds, and I fortunately met and dated a few, flattered to be with each one. Those times were exhilarating, full of joy. However, my locomotion of faithful devotion had earlier been derailed, and I had little interest in hopping back onto the railroad tracks of true romantic commitment. I was distant and hesitant of another deeply serious relationship, and it was my loss. Those charming girls, especially one, deserved much better.

Though I saw Ed only a few times, I occasionally escorted dates to his chapter house for all-Greeks-invited weekend dances, identical to the scene in *Animal House* of the fictitious all-black R&B band Otis Day and the Knights joyfully entertaining exuberant and intoxicated all-white toga-clad couples with 1960s "beach music." I feel confident that most elements in his chapter's environment were the same as in mine.

In the spring of 1976, Ed graduated with a degree in business. During his four years, he and the other managers under head coach Bill Battle prepared the football team to achieve a most respectable record of 32 wins, 14 losses, and 2 ties. I knew nothing of his coursework, but it surely included statistics, real estate, marketing, management, insurance, and other topics that were like distant languages to me but ones that would provide a strong foundation

U.T. graduation, 1976   UT

for his future. When he departed the university with a diploma in hand, Ed received the deserved gratitude from the Vol players that he and his squad had helped train and kept in excellent shape during the past four football seasons.

My experiences in the five-year architecture program were quite different from Ed's, with long hours formulating and testing creative building designs and then preparing large drawings for the often-brutal evaluations from professors at the exciting, still-new School of Architecture. Students were allowed to choose professors for the arduous design classes, the most important one that required full dedication and unending hours of work. Though most of these teachers were admired and always selected, instructor Bill Shell had a terrifying reputation. Since no one signed up for his class, a lottery was held, and I joined ten anxious others who received the short end of the deal. Professor Shell was simply demanding and wanted nothing but excellence, and I survived his class and received an A, the proudest grade in my five years at UT. Other influential teachers included Sinclair Hui, Duane Grieve, and especially Peter Lizon. Lizon pushed students to enter national design competitions arranged for students in architecture, and a few of us placed in those prestigious contests of creativity.

Architectural student

I lived off campus my final fifth year in the Townview Terrace apartment complex, and I worked part-time at the architectural branch of the Tennessee Valley Authority. Away from university

Second Largest Graduating Class at UT

U.T. graduation, 1977 UT

dining halls and my fraternity's wonderful cooks, I quickly learned that my skills as a bachelor chef were less than adequate, so I had to rely on another chef named Boyardee for his canned pasta specialties. I also survived on chicken pot pies, fish sticks, and other frozen delicacies. One year after Ed graduated, I walked the same path and joined another gigantic "host of Volunteers" on June 10, 1977, donned the black graduation gown, marched into the massive Stokely gymnasium, and seized the graduating ticket to the future. With 2,307 graduates, it was the second largest in UT history. As at that final ceremony seemingly eons earlier at Central High, everything would again change forever. Ed was back in the Boro, and I imagine that he and his siblings purchased Dan Fogelberg's newly-released album. Its title of *Nether Lands* was derived from the quiet town of Nederland, Colorado, where part of the record had been recorded and one that looked, on a postcard, to be idyllic, soothing, and inspirational.

Nederland, Colorado

Fogelberg's album SME

CHAPTER 7
Next Adventure

The fall after I graduated, I accepted a position in Nashville as an intern architect with the rapidly-growing architectural firm of Gresham & Smith. It was known for its successful relationship with another, quickly-growing entity, Hospital Corporation of America. HCA kept G&S busy designing for-profit hospitals for locations across the nation. With over 100 employees, the firm was a great place to start, and I received much-needed practical experience under the guidance of Jim Clement, Tom Bulla, and Jim Wachtel. I also started studying for the dreaded architectural licensing exam. Passing it after the required time in an architectural firm would be all important, and a group of fellow employees and ones from other offices formed to study and ensure that we would all become licensed.

Due to encouragement from her close friend, I finally called the attractive blonde that I had seen at that Carrick Hall bus stop during my freshman year. Like Ed, she had graduated after four years at UT, and we were both single and unattached. I was so nervous when I phoned to ask her out, I had written topics on a notecard to keep me from fumbling during the conversation. She was one of the most delightful girls on the entire UT campus, and I was overjoyed that this popular and very smart young woman would agree to step out with me.

With the exception of hearing that Ed had joined his father's insurance firm, I knew little of what he did immediately after college. He and other young bucks in the Boro had a blast

Gracious Ed with friends

as boisterous ones, working and playing hard as bachelors. They played tricks on one another, and one evening, unbeknownst to Ed, fellows out for fun saw his distinctive convertible parked at a bar on East Main. Silently they put the car in neutral gear, and then quietly pushed it around to the back of the joint. Discovering that his prized vehicle was missing, furious Ed called the police, and when two patrol cars arrived with flashing lights, it scared the pants off the pranksters who watched from a distance as a formal theft report was filed. They remained silent about their now-serious practical joke gone awry. Later when one of the culprit's equally cherished vehicles, a 1963 Impala SS, went missing late on a weekend night, the owner knew it was revenge and called Ed in the wee morning hours demanding to know where his car was now hidden. Somehow, Ed knew of the person's involvement in "relocating" his convertible, but he wasn't involved in this prank. Perhaps enjoying the moment, he denied any participation or info on the muscle car now gone. The second vehicle had actually been stolen by a black couple who drove the fast machine to Nashville, wrecked it, then fled the scene and were never apprehended. The car was finally retrieved by the hacked owner from the Metro police tow-in lot.

While Ed's business with his dad increased, I passed the rigorous exam and became a registered architect. Yet desiring to obtain additional design skills and see the Northeast, I applied and was accepted in the Masters of Architecture program at the University of Pennsylvania. Once in Philadelphia, I quickly learned that my accent was not the least bit cosmopolitan nor neutral and was instead distinctly obvious. I was forced to tolerate frequent misconceptions about Tennessee and the South. Thus, I became friends at Penn with students from many other foreign lands including Japan, Brazil, Greece, and South Carolina. Looking back, I was likely admitted for diversity as the token white boy from below the Mason-Dixon Line. Regardless, graduate school at the Ivy League institution was exhilarating, and I was invited to attend business classes at its prestigious business entity, Wharton, where Ed's older brother had studied many years earlier. I was also honored to work part-time in the office of Venturi, Rauch, & Scott Brown, one of the most influential design firms in the nation. The conclusion of my coursework took place in Paris at the distinguished Ecole des Beaux-Arts. The life-changing summer included living in architect Le Corbusier's celebrated 1931 Pavillon Suisse dormitory, tours of every important building in France, and being completely mesmerized by brilliant, thundering fireworks encircling the Eiffel Tower on July 14, Bastille Day. I received the advanced diploma and quickly headed back to Nashville for the lovely and thankfully patient blonde.

Before we dated, this girl had graduated from Tennessee with a math major. She immediately achieved her lifelong goal by purchasing a car, renting an apartment, and then buying a thoroughbred horse and a horse trailer. Yes, she was one of those little girls who had grown up completely fascinated with the four-legged creatures. She would rather clean a stall than clean her room. It was crystal clear from the beginning that in our relationship and any possible future, I would know that large, pasture-requiring, land-damaging, hay-eating, expensive-feed-devouring, wintertime-barn-occupying, and manure-only-producing creatures would always come first. I had wisely returned home for the girl, immediately asked for her hand, and a few months later introduced my fiancée to classmates at Central's tenth year reunion. Wisely scheduled back in 1972, the reunion held in the large ballroom at the Ramada Inn on December 26, 1981, was a gigantic success. The prediction that so many graduates would be back in the Boro proved beyond expectations with overwhelming attendance, and rambling through the packed crowd of attendees, all of us saw what ten years had done to each other. With my girl, I was in heaven. Marilyn and I tied the knot three months later.

Central High 10th reunion   RB

Four years then went by. One day, I passed Ed on a sidewalk that encircled the town square in the Boro. He had reportedly become quite successful in his dad's long-established insurance company. The respected agency owned a prominent structure built in 1900

NETHER LAND

at the corner of North Maple and West Main, one of many that preserved the charming ambiance of historic commercial buildings surrounding and visually exalting the landmark 1859 Rutherford County Courthouse. True to his desire for prominence, giant letters NETHERLAND COMPANY were painted on the main, east-facing façade at this 101 North Maple address, leaving no doubt about the talent and egos inside.

A year before seeing Ed on the pavement, I had started a new architectural firm in Nashville with Auburn graduate Seab Tuck and UT graduate Gary Everton. Due to the name of our studio, Tuck-Hinton-Everton Architects, we paid for granite cornerstones to be installed on constructed projects, each engraved with the year completed and its design by "T.H.E. Architects." It was admittedly obnoxious, yet the three of us had lofty goals and wanted to make our mark. I was now trying to obtain clients in my hometown.

101 North Maple Street   RB

Ed and I exchanged greetings at that brief encounter, yet his insincere smile and air of superiority immediately reminded me of events a decade earlier. Although we had led superficially parallel lives up to that point, I knew very little about him and had no interest in his activities in the Boro. Yet as I walked on, I wondered if in the course of obtaining wealth he had cultivated others who disliked him or were serious enemies. My underestimation of his capacity to do this would be enormous. Perhaps it would lead one day to his murder on a faraway island.

CHAPTER 8
Overture to Sagas

My research about Ed is because someone had stolen one of the numerous items on my bucket list. I assume everyone has such a ledger, and throughout my life I've slowly marked through satisfied encounters, accomplishments, destinations, and obligations. Two were marked off when I traveled to Egypt to see—and enter—the enormous Khufu pyramid, and when I ventured to Athens to see the Parthenon. Perhaps Ed had his own bucket list. Raised in similar circumstances and educated at the same institutions, our lives after college took wildly different routes, mine traveling slowly upward on a smooth country road and his careening along an unlimited-speed autobahn. After our vehicles collided at his earthly sendoff, one item on my list vanished. Ed's tragic death denied me a chance to accomplish it, and in its place, I simply had many questions needing many answers.

A smart and gifted businessman, Ed quickly became very successful in his insurance company. He was gregarious and aggressive, and married a prominent, younger Murfreesboro ingenue in 1980. The marriage failed a few years later. Ed was focused on his business, wanted to get far ahead, had his eye on fancy cars, and was always thinking of the "home run" deal. He would prefer to insure not a person or company but instead a large institution or huge corporation. In the coming decades, he would succeed. But his journey on the avenue of success would be a bumpy ride, upward and downward, dodging regulatory roadblocks, sailing quickly through banking industry toll booths, and exceeding the speed limit while scooping up coins on business highways paved with shiny silver dollars.

The path to prosperity that Ed selected was an excellent fit for his personality and his ingenuity. His profession of peddling life insurance was a necessary yet perplexing safeguard requiring a person to gamble on his or her own life for the protection and benefit of family or others. An industry full of its own lexicon, it includes well-known terms such as premium, which strangely as an adjective means "higher value" yet as a noun means "shelling out cash for coverage." Ed studied the rules and actuarial statistics, and he developed several untested approaches on how premiums could be gathered and death benefits paid. Evaluating tax codes, he also understood maneuvers that, on a large scale, could make him extremely wealthy. Though based in the Boro, his insurance company was listed in the Nashville phone book, communicating his availability in the nearby large city. Eventually, he would become a noted pioneer in a controversial insurance technique known ironically by the acronym of a purple flower. Its intoxicating payback for some would be balanced for others by the manmade plant's sharp and painful thorns.

In contrast, my career was a slow process involving convincing potential clients of my capabilities and our firm to deliver an appropriate design fulfilling their needs and dreams for a new house, school, library, apartment complex, restaurant, or office building. Only with years of success in modest projects would clients entrust our firm with larger, more complicated, and more prominent commissions, most notably to design a new church or museum. Even with the most distinguished assignments, the fees would be modest. Thus, architecture as a trade rarely guaranteed the higher profitability of many other businesses or professions. The main payback was the satisfaction of serving clients and the privilege to conceive something appropriate, unique, and perhaps inspirational. Architecture was also my hobby, a jealous mistress providing sheer delight while requiring constant, serious devotion to her demanding craft.

Though Ed was making good money from the insurance business, he always kept his eye on other ways to make dough. He wanted more. He was living in a lovely house on East Main Street, only a block east from the Central High building which had been converted into a large junior high school. He was brilliant in learning alternative ways to increase his already sizeable bank account. Yet before enormous successes inventing groundbreaking insurance techniques, Ed tried to make a blooming fortune in a completely unrelated business. Twice he attempted to plant a manmade creation, one that might become an industrial garden producing enormous crops of green dollar bills, yet one that might also defoliate its surroundings in a toxic heap of physical, environmental, and humanitarian damage. Just as my son was born in the spring of 1989, Ed would soon birth his technological creature, and he would nourish it like any devoted father.

CHAPTER 9
Trash Endeavor One

Why Ed decided on a different way to make money is unknown. He turned his effort and energy from the business of coverage to the business of trash, really bad trash. At two different locations, he had the same objective—raise money to receive local and state approvals and then finance construction of a toxic waste incinerator. Such a facility was in great need, yet no community was interested in providing the land and infrastructure necessary for its establishment. However, given the potentially sizzling financial rewards, Ed went after it and twice pursued the incinerator idea, first in Tennessee and then in Mississippi.

Giles County, Tennessee

With others in the summer of 1989, Ed formed FederateTechnologies Incorporated, a venture capital company that would attempt to build a lucrative hazardous waste incinerator somewhere in rural Tennessee. This incinerator would burn biological, chemical, and other dangerous garbage, refuse that simply couldn't be tossed into a landfill. Among initial investors in this challenging endeavor were several deep pockets, including well-known Nashville restaurateur, Ray Danner. After looking at several locations, the first place Ed seriously considered was Giles County, a picturesque 611-square-mile rectangular province with its center the quaint, historic town of Pulaski. This was the community where the notorious and secretive Ku Klux Klan may have started immediately after the Civil War, with this group's initial mission to harass crooked carpetbaggers from the victorious North and, later, to terrorize freed slaves who were attempting to establish themselves in the confusing post-war society of Reconstruction. Now, FTI would develop a comprehensive battle plan and, not unlike events 123 years earlier, mount an assault.

Giles County courthouse

Ed purchased a parcel of land including a house in a rural part of the county, intending to erect the large "job-creating facility." Though extremely contentious, Ed's project was a needed facility, one to deal with scary refuse and particularly dangerous garbage that had nowhere else to go. Yet almost immediately, local and state officials expressed concerns and opposition to his plan, and as spring became summer 1990 in Giles County, the souring humidity around the FTI idea started to soar. Ed was discreetly warned by friends from Murfreesboro with dependable connections in Pulaski that fear of the proposed waste operation was gaining serious momentum. To complicate his personal situation, Ed had been earlier diagnosed with melanoma, and with a new wife and young children, the 35-year-old was now under mounting pressure to succeed with his controversial project.

Downtown Pulaski

Though Giles County was the scene of only one significant event in the Civil War, the Battle of Anthony's Hill in December of 1864, it might have been wise for Ed to appreciate that confrontation and recognize the impending one he was facing with a similarly determined protector of the lovely countryside named Carol Puckett. Raised in Decatur, Alabama, Puckett had always admired the pastoral quality of this distinctly rural Tennessee county. She attended the University of Alabama in the early 1970s, majoring in journalism and minoring in the emerging study of the environment. She married a computer software designer who possessed the same gregarious nature as she did, and they eventually moved to Giles County in 1989. A privileged white lady with authentic roots in the Deep South, Puckett possessed a candid, lively, and intelligent manner and would never be pushed by those who thought she should be genteel and compliant. Learning about the somewhat mysterious FTI incinerator, she used her journalism training to gather detailed information on the company and its founder. Discovering the potential harm of this industrial proposal and observing that it was quietly moving forward, she quickly helped form a local group of opposition and gave it a name that was downright crystal clear, "Citizens Against Toxic Incineration."

Puckett was a fiercely determined lady, equal in strategic thinking and moves that might have impressed the two commanders of the Anthony Hill confrontation, Confederate General Nathan Bedford Forrest and Union General James Wilson. In late 1864, the rebel troops under Forrest were retreating from disastrous losses in Franklin and Nashville when they made a gallant stand in Giles County. But Wilson's Union men carried on and drove Forrest and his troops westward to the Tennessee River. In this county and others in the South, the Civil War would later be called "The War Between the States," "The War of Northern Aggression," and a subtle favorite, "That Late Unpleasantness." Under their determined leader, the CATI group and its many supporters were going to make sure that Ed and his project, like the rebels 126 years earlier, would be pushed out of their county and hopefully run out of the entire state. They met and discussed every conceivable piece of political, intellectual, environmental, and emotional ammunition, and prepared to fire at FTI's team and its proposed incinerator.

On July 5, Ed explained the concept for the proposed facility to a group of local businessmen during that most effective sit-down marketing maneuver, a free lunch. Yet most quietly listened to the sharply-dressed salesman's initial remarks and then reportedly left without eating. Regardless, Ed remained confident and felt that he was not a dastardly hustler attempting to bring an actual, physical monster to Giles County. He knew that the proposed facility was necessary and had to be established somewhere, so why not there? He also knew of the profitability of something unpleasant yet necessary, and his swaggering courage and detailed strategy had been impressive to those who tossed large dollars into his can for this venture adventure. Their investments would sadly not germinate in this waste bucket.

The local rag *Pulaski Citizen* issued a special bulletin on July 17 about FTI creature, and it was becoming evident that almost no one wanted Ed's plant to be planted in Giles County.

Ad in Pulaski Citizen   TPC

Just a few miles north of Pulaski was historic Milky Way Farm, a scenic agricultural place founded in 1930 by the owner of Mars Candies and famous for its 1,100 acres of cultivated

Milky Way Farm   RDS

property and distinctive agricultural buildings. Its sweetness representing the community was appreciated, and it was featured on county promotions and visitor post cards. Ed's potential incinerator would never, ever receive such flattering coverage.

Public opposition to the FTI endeavor became seriously evident during a protest rally at the Pulaski town square on July 21, 1990, one arranged by CATI and attended by the largest crowd in the county's history. Traveling south from the capital in Nashville a few days later on July 24, keynote speaker Governor Ned McWherter informed a cheering group at the local National Guard Armory that the toxic incinerator would not be allowed in this or any other county in Tennessee. He signed the CATI petition against the project. The unwanted plant received fatal, governor-infused herbicide and was now a big dead weed.

Protest rally in Pulaski   TPC

Regardless, CATI officials wanted to be sure that Ed and his FTI army were absolutely crushed. This was still a serious war, and before the governor delivered political poison to kill the toxic plant, the group had decided to launch a very personal attack aimed at FTI's commanding officer. It was as if the Civil War had returned when the group loaded their cannon and fired from Pulaski a gigantic artillery shell sixty miles over Marshall County to land in Rutherford County at the front office of *The Daily News Journal*, Murfreesboro's daily newspaper. The shell was packed not

Mr. Ed Netherland has stated his intention of putting a toxic waste incinerator like the one pictured above two miles as the wind blows from the Giles County Court House and less than 0.25 miles from our water source.
Mr. Netherland has stated that a "small but vocal group" of environmentalists oppose this

Ad in Daily News Journal   *RCH*

with gunpowder but a fat check, as the CATI supporters paid for a large print ad, one with its bullseye on Ed's forehead. Knowing that this piece might ruin any of his daily encounters with the homefolk on his home turf, they pulled off all gloves and directly accused the native son of full responsibility for the terrifying incinerator, an industrial plant that was going to poison pristine Giles County. The large printed protest ran in the Murfreesboro paper on Thursday, July 26, 1990, and residents in Pulaski could hear the boom far away in Rutherford County as citizens opened and read the ad. Then they saw the political smoke rise.

The CATI ad also urged attendance at a rally two days later in Nashville at War Memorial Plaza, the spacious civic setting directly south of the Tennessee State Capitol. Since its completion in 1974, this urban plaza provided exceptional views of the landmark Greek Revival statehouse designed in 1845 by famed Philadelphia architect William Strickland, and the majority of protest events aimed at Tennessee's elected officials were held in this large, granite-paved open space. Death nails had already been hammered into the FTI coffin by the governor, and the abusive ad and rally weren't really needed. Regardless, this unprecedented personal assault on Ed was a direct plea to those who knew him or knew of him to grab the scallywag, remind him of the probable irreversible environmental damage his plant might deliver anywhere, and if that didn't work, publicly embarrass his family, squash his ego, and completely stomp the fire out of his infernal project.

With the governor's statement, solid opposition by officials and influential citizens, and personal attacks, the FTI endeavor in Giles County was wadded up and tossed into the round file. Several FTI board members resigned. Ray Danner, embarrassed by the tremendous backlash and boycotting of his Shoney's restaurants, withdrew his coins and public support, yet he was well aware that the smart guy from the Boro was formidable and worth keeping on his Rolodex file for future opportunities. I learned that Ed was also informed in a phone call that the insurance on his property's vacant house in Giles County was being cancelled, and that the structure might soon be burned to the ground as a subtle, country-boy warning for him to skedaddle.

CHAPTER 10

Trash Endeavor Two

Ed may have been adventurous, but he wasn't stupid. He had investigated many other possible sites for the FTI project, knowing that the incinerator was still needed somewhere below the Mason-Dixon line. Since Giles County and Tennessee's governor had turned their thumbs down, he would simply plant his plant elsewhere. Ed slid his red-hot proposal three counties west and then straight southward across the Tennessee-Mississippi state line five more counties to an impoverished and possibly welcoming locale. Based on thorough research and knowledge of the sad shape of Mississippi, poorest state in the nation, he now rested his ambitions in Noxubee County, an almost perfectly square, 695-square-mile entity on the east central part of the state.

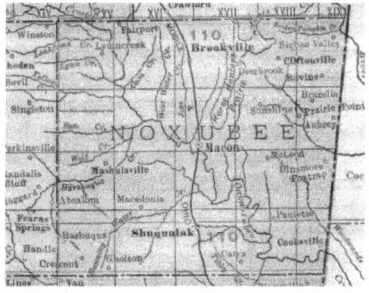

Noxubee County, Mississippi

Compared to Giles, Noxubee County had one-third the population, more serious rural poverty, rampant illiteracy, and long-standing racial tensions. Based on median income, it was near the very bottom in this extremely poor, Southern state, making

Noxubee perhaps the most economically depressed and desperate place in the entire nation. It was screaming for jobs, and Ed and his team at FTI were ready to "help" by trying to establish here their solution: that same employment-creating, hazardous toxic waste incinerator. The citizens of Mississippi did not relish that their state was at the very bottom of many national evaluations of education, income, and racial harmony, and consequently almost any endeavor that could deliver paychecks was worth pondering. Waving the word "jobs," regardless of the attached details, worked especially well in this sad place.

It didn't help that Noxubee County's name had been derived from the Native American word "nakshobi," which means foul-smelling water. Yet the place had many picturesque landscapes, including forests, fields, and lovely streams. A modest brook was named Dancing Rabbit Creek, and it had a special designation and heritage to go along with its smooth, gentle, and perhaps healing waters.

Ed had an office in the Boro, but he needed a more convenient location for his FTI team to meet and strategize with experts, legal minds, and public relations gurus. His new setup in 1991 was inside a recently-constructed office building at 1800 Church Street in Nashville. Here his FTI team would contemplate and debate in prosperous Music City the chess moves to make 300 miles away in the faraway, completely unprosperous Noxubee County. Coincidentally, my architectural firm was only one block away, immediately south at 1810 Hayes Street in a renovated, historic Victorian Style residence built in 1910. Although there were numerous places where Ed's paths and mine might have crossed—especially during lunchtime at the numerous nearby restaurants including Randy Rayburn's splendid Midtown Cafe—such never occurred. Similar to days in college, our worlds and activities were quite different, and our business and social planets never aligned. I had absolutely no idea of his toxic waste adventures in Tennessee and Mississippi.

Racial tensions in Giles County during Ed's time of stirring up environmental dirt were minimal, but not in Noxubee, a county in the deep, Deep South. Rarely did Caucasians and African Americans ever engage, and the public schools reflected this condition. Almost all white students attended private schools while students of color comprised 96.5% of the public-school enrollment, a stunning example of noble efforts for educational equality gone completely wrong in many parts of the South. Mennonites also comprised an appreciable amount of the population, yet they puzzled many by not participating in local politics, not voting, and not showing real interest in others. Noxubeeans often joked that their county was composed of three completely separate races: whites, blacks, and Mennonites. Native Americans inhabited the county yet were rarely included in that assessment.

Noxubee County courthouse

CHAPTER 11
A Past to Remember

Ed's reconnaissance of Noxubee County likely identified a prior effort in the 1980s to capitalize on the region's poverty and desperate unemployment situation. But he may not have fully understood its comprehensive history including not only the acknowledged deplorable period of slavery but the equally disgusting saga of the area's original inhabitants, the Choctaws. Although comprising a small percentage of the county's population, these Native Americans would not be disregarded. Transgressions long ago would never be forgotten, lasting as permanent three-

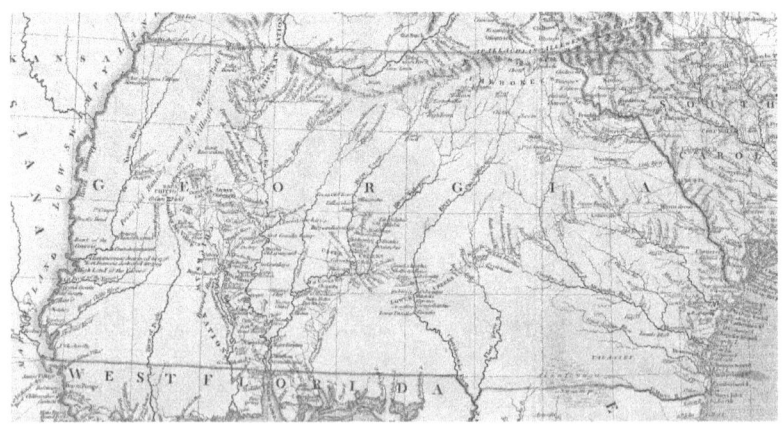

Map of Georgia, 1806

way distrust among Caucasians, Native Americans, and African Americans. It was a circular firing squad of suspicion.

The state of Mississippi possessed a unique heritage which started with one of the original thirteen colonies. Georgia initially existed as a mammoth entity just below Tennessee, stretching from the Atlantic Ocean all the way to the Mississippi River. The Choctaw people resided and hunted on land in the southwest corner of what would become the Deep South, and though many treaties had been signed between the native people and the arriving whites, most had been ignored or completely dismissed, not worth the faded ink signatures on buffalo hides. With the Choctaw, this would fortunately not be the case.

Choctaw Native Americans

Congress divided the enormous original colony of agriculturally-rich Georgia into three, almost-equal rectangular slices, and the two western ones became official territories named Alabama and Mississippi. In 1817, the one bordering the grand Mississippi River became the twentieth state. A similar event had occurred in 1796 when the long parallelogram of Tennessee had been severed from North Carolina. Europeans and those from the Eastern states had been establishing large plantations in the South where the soil was extremely rich, and their huge farms of cotton and other valuable agricultural products would be planted, tilled, nurtured, and harvested by free labor. A cynical historian might write that the leaders in the rapidly-evolving Southern states seemed to have three goals: first, grab all land and establish permanent territories for the expanding white population; second, get rid of most if not all of the indigenous tribes that claimed these areas as their homelands; and third, ship in even more captured slaves to cultivate the land. Property and slave owners would become prosperous beyond belief, creating a strong white society of privilege, education, entertainment, and the finer things in life. Meanwhile, enslaved blacks were treated like livestock, and the rights of native tribes to their homelands almost completely vanished.

The final blow in the dispute over the soil of Mississippi and other states came with the Indian Removal Act signed into law in 1830 by President Andrew Jackson, mandating the relocation of most native people to territories west of the Mississippi river. Regarding the Choctaw, their removal came with an agreement for them to vacate 11 million acres in exchange for 15 million acres far away in what would later become the state of Oklahoma.

However, there was a unique component in a separate agreement, the Treaty of Dancing Rabbit Creek. The document was signed on February 24, 1831, by Choctaw leaders and representatives of the federal government at a spot in the southwest corner of Noxubee County. This agreement allowed Choctaw who wanted to remain on their ancestral lands to stay put. The few that decided not to leave were granted reservation property in what was the first major recognition of a non-European ethnic group, allowing these

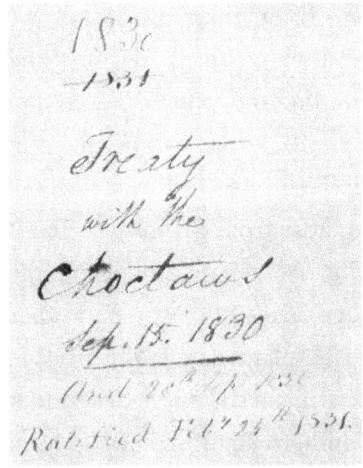
Treaty of Dancing Rabbit Creek

original inhabitants to become American citizens. But slavery of fellow human beings in the South continued to cast a dark shadow over the nation, especially in Mississippi.

Slaves in Mississippi

CHAPTER 12
Making it Big in Mississippi

After ridding itself of most of the Native American tribes, the taste for slavery in the still-new United States continued at an aggressive pace. The prosperous state of Mississippi would eventually have more humans without freedom than those with. Sadly by 1860, 430,000 blacks under the whip unjustly enriched the state's white population of 350,000. Wealth in the South grew to unimaginable levels. The city of Natchez on the Mississippi River had scores of millionaires and perhaps the richest per capita (white) population anywhere in the Northern Hemisphere. Even after the tragedy of the Civil War, freed slaves would forever struggle to obtain an equal status, joining the Choctaw as citizens with limited rights and lackluster opportunities. Noxubee would never be on a postcard of racial harmony nor of fairly-distributed economic prosperity.

Ed may not have been fully aware of the complicated history of the inhabitants, yet he definitely knew that other

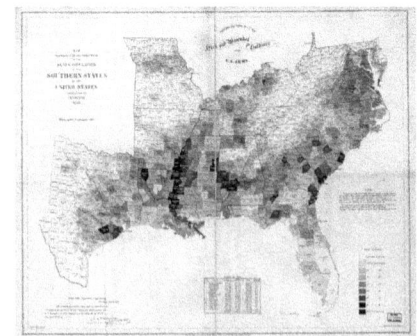
Slaves in the South

garbage companies had their sights on Noxubee. Seven years earlier in 1983, a serious confrontation had occurred between county citizens and Chem Waste, a subsidiary of the enormous garbage company, Waste Management Incorporated. This trash entity ran into fierce opposition from several strong-willed women who, along with others adamantly opposed to any new garbage establishment, chartered a nonprofit to raise money and raise outrage. It had a catchy nickname, and their organization Protect the Environment of Noxubee would be known as "PEON," a direct jab to those in power who might perceive these folks as weak.

It was a divisive struggle, and in the end, the opposition temporarily triumphed in 1985 when state officials announced a moratorium of five and one-half years for any commercial waste facility. Both sides of the fight knew that when the sixty-six-month ceasefire ended in 1990, the conflict would probably start all over again. Newcomer Ed was therefore stepping onto a battleground

Waste Management dumpster

populated with the injured veterans of the previous trash war, one still strewn with emotional shrapnel of distrust, hate, and greed. He would face opposition from citizens of different races and also from patiently-waiting Chem Waste.

CHAPTER 13
Planting The Plant

Ed had learned a great deal from the failed incinerator attempt in Tennessee, and he recognized that too many in Giles County resented his secretive dealings and back-slapping, good old boy tactics. This time he would be more strategic. He would be open and candid, immediately contacting and courting as many locals as he could. He established a crucial relationship with Brad Moore, a county official and mayor of Brooksville, a small village of 1,000

Macon, Mississippi   CR

just nine miles north of Noxubee's County seat in Macon, an equally quiet town of 10,000. Moore filled many categories that Ed needed for support. A real estate agent with deep generational roots, he was keenly aware of the politics and people all over the county. He became a key supporter and stood behind Ed against the opposition, and the two became best friends in the contest.

Both Ed and Moore hoped that they might develop with other influential businessmen a home court advantage, something extremely valuable in this Deep South place of tradition and established rules. Moore suggested that Ed enlist the help of Ben Tubb, an admired businessman who had been recognized as "Man of the Year" by the county's Chamber of Commerce. Ed and Tubb met and agreed on a contract for this life-long Noxubeean to provide local public relations advice. He completely understood the community-engagement errors Ed had made in Tennessee, and he recommended that scores of meetings needed to be held with any and all of influence to politely introduce FTI and its star promoter. These sessions would purposely have a limited number invited each time and would occur all over the county in eateries, taverns, social clubs, "juke joints," churches, parks, schools, and at the all-white private country club.

Another supporter of the FTI proposal was Ralph Higginbotham, head of the county's five-member Board of Supervisors. Raised in the hilly southwestern part of the county, he came from a poor white family, had limited formal education, and was a proud and stubborn man with suspicions about the power of those who had inherited generational wealth. With this tough one on his side, Ed was seen as the friend of poor whites.

Tubb and Moore stressed that Ed needed to meet and greet

PLANTING THE PLANT

African-American citizens in the same manner as he did with whites. They introduced Ed to activist Ike Brown, an outspoken, jovial black man who had influence in the county, especially with members of the local chapter of the National Association for the Advancement of Colored People, the NAACP. Brown provided excellent advice and helped Ed meet with black citizens who would be involved in the contested waste proposals. Many were descendants of slaves who now simply wanted and desperately needed any situation that might provide stable employment.

Brown would become Ed's ace up his sleeve, totally supporting the Tennessean's FTI proposal. He believed that white guy Ed was completely devoted to improving the economy of Noxubee and was helpful, polite, intelligent, and devoid of racial prejudice. Ed would help those of color who had the indisputable and solid 70% majority of the county population and now sought long overdue financial prosperity and authentic political power. This would not be the first or last time that a solution to obtain decent jobs would be unsavory, and yet many of these residents rightfully feared that if the FTI facility or another waste version was not established, no other paycheck-providing plant would ever germinate from the barren business dirt of Noxubee. Activist Brown worried that the county's black populace would then simply die from poverty.

Ed's hazardous waste facility needed a good site, and Moore helped him find the perfect dirt. Although FTI needed approximately five-hundred acres, they found an enormous available parcel of six-thousand acres on U.S. Highway 45, the main road connecting Macon and Brooksville. The land had been purchased four decades earlier by a lawyer way up north in Indianapolis, and this benevolent man and his wife donated the property to the Indiana

University Foundation. The Northern institution had little use for land so far away and was ready to unload it, yet there was a stipulation that if ever offered, the property must be purchased as a single parcel. This gift had been appreciated, but it would really be appreciated when sold, thus allowing an armored car full of Southern cash to make the 600-mile trek straight north back into Yankee country to its final destination, the University Foundation's boardroom in Bloomington.

Ed informed his investors that this gigantic tract of land, priced at a steep $3.5 million, might now allow FTI to add to the already big incinerator creature and establish an adjacent permanent hazardous waste repository, a profitable dump for what one anonymous trash disposal official called, "da baddest of all bad stuff." He received approval from his financial backers to purchase an option, received thanks from hopeful Foundation board members far away, and announced with undeterred enthusiasm that a new and improved, comprehensive waste facility would provide even more employment for not only Noxubee but surrounding counties, including those in Alabama across the nearby state line. With this megalomanic, full-service disposal center, Ed and his massive facility would become famous, perhaps one day to be featured on Mississippi's economic incentive brochures, although probably not on tourism pamphlets or postcards.

CHAPTER 14
Trashy Competition

Making Ed's challenge even more complicated, other waste companies aimed their gun sights on Noxubee. Chem Waste had been waiting since 1985 and was ready to go. With existing connections and scars from the earlier conflict, they had a long history in the region with sour accusations over their other garbage facilities. They were seen by some as a bad actor with a bad record, whereas Ed was the new kid in town with an unknown, untarnished reputation.

Ed would also have competition possessing a Choctaw twist, National Disposal Systems. Its owner was a knowledgeable one from Jackson with valuable experience in waste disposal and chemical cleanups. In 1988, he had approached a controversial Choctaw chief with an ingenious idea that would involve the leader's people. National Disposal Systems would purchase land somewhere in the county, donate it to the native tribe, receive a long-term lease, and then establish a profitable disposal dump on the rented parcel. The purchased land would not be on traditional Choctaw soil but on property now owned by them, thus using Native American sovereignty to dodge state agency interference. It was a darn clever idea, not unlike the revenge all tribes throughout the United States would achieve with casinos on their untouchable reservations.

Even though the FTI waste management facility had not received any approvals, Ed and his company made another smart move by making job applications to work at the plant readily available, and scores of locals submitted their immediate interest and availability. To avoid the impression that his numerous smaller meetings were secretive, Ed printed an open letter in a local shopping guide that was gracious and yet clearly conveyed that public support was crucial if FTI was to select Noxubee as their final choice. Playing "hard to get" was effective, as those supporting the project now became more vocal, sharing their enthusiasm for this technological facility and reminding those sitting on the fence that the project might be established somewhere else in Mississippi, once again leaving this sad county in the dust.

At first, the promise of this new industry created excitement, enthusiasm, and broad support. To Ed's delight, his supporter Brown soon convinced the NAACP chapter to formally endorse the FTI scheme. With the assistance of individuals who were consumed with finding any jobs, Ed mounted a full-court press to receive governmental approval. Though his many presentations had been in the open, he was shrewd and used every political maneuver, including backroom discussions with local and state regulators and with any elected official he could find.

In Giles County, Ed had encountered the fierce female opponent, Carol Puckett, and his cruising FTI dump truck slammed into yet another wise and determined woman in Noxubee County, long-time resident Martha Blackwell. The folks who fought and won a ceasefire back in 1985 were still around, and now the PEON organization needed a five-star general to triumph over this returning nightmare. Blackwell had become aware of the FTI toxic

TRASHY COMPETITION

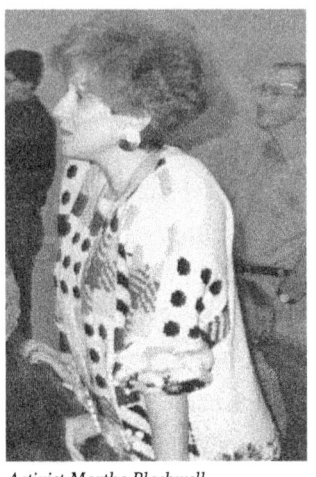
Activist Martha Blackwell

waste endeavor in late 1990, and she quickly rose to the occasion. She was a wealthy, college-educated white woman whose ancestors were among the first white settlers to farm Noxubee County. There had been no Civil War battles in this place, and yet the struggle could have matched ones elsewhere in Mississippi during that terrible national conflict. She and PEON were now waging war against not just rascal Ed but other men of garbage.

It would become evident to Ed and others that the landfill proposal from National Disposal Systems was going nowhere and was headed for the round file. The Choctaw chief promoting this scheme thought that profits from the National Disposal Systems scheme would enrich his people, but grumbling local and national tribal members adamantly stood against dumping of trash on their land, ancestral or recently acquired, nor upon any other location in their county. The seemingly untouchable proposal that would skirt around state and federal authorities would itself eventually go into the waste basket.

Ed knew that competitor Chem Waste also had its own problems. Though having credible regional experience in the landfill business and involved in the waste tussle of five years earlier, they nevertheless had a mixed reputation. Citizens from nearby counties familiar with Chem Waste disposal facilities spoke out, adamantly trashed the trash company, and warned others that all they had received was a massive cat box. Ed had also worked harder to gain strong support from the diverse population. The folks with Chem Waste never fully matched his skillful marketing nor would overcome the early endorsements bestowed upon FTI. Compared to Chem Waste's smelly dumps elsewhere, Ed's untested incinerator scheme seemed to be a breath of fresh air.

The end of 1990 was approaching, and the competing waste companies needed a decision by the county Board of Supervisors. Ed convinced Higginbotham that the board should go with FTI and adopt a resolution limiting the competition, but this drew the ire of other officials, and after heated discussions, the board decided on November 21, 1990, that there were too many questions and a single endorsement was unwise. Yet only a few days later, Higginbotham seconded the motion by a black board member to endorse FTI's proposal alone, and this time it passed three-to-two. Without vital information and clarity, this vote came as a shock, especially to Macon attorney Timothy Gowan, the board's legal advisor.

Gowan was suspicious of the financial documents provided by Ed. The FTI waste company was incorporated far away in Tennessee and had few tangible assets, and he worried about its true financial stamina to build the very ambitious waste facility. Gowan also asked Ed if he would indemnify the supervisory board and the county if anything went wrong, but no forms were signed. The Macon attorney believed that money was being spread around not only above but beneath dining room, coffee, and picnic tables to influence or even purchase support. Reportedly, he felt the whole thing was a farce.

With the endorsements, Ed and his FTI proposal now seemed to have the upper hand. But the required approval from state authorities had not arrived, and what would become his company's most serious competitor entered the crowded waste war in March 1991. Headquartered in Houston, United States Pollution Control Incorporated was one of the largest trash handlers in the nation. Due to its sheer size, unmatched experience, and solid reputation in the dirty, unpleasant, and often dangerous garbage business, USPCI immediately became a threat to the other proposals. Unlike the mysterious financial documents and promises from Ed about FTI, this national contender's balance sheets made heads throughout Noxubee spin wildly. The stout, fully capable corporate giant was late to the game, entering the fight several months after the county's earlier endorsement of the FTI proposal. But they depended on existing and refurbished political connections with many of the most fortunate and wealthy in the county, and they named a native son to shepherd their disposal proposal. Ed was charming but still an outsider, and the USPCI folks were seen as polite and honorable ones representing the established aristocracy.

Now it seemed that the waste war was between FTI and USPCI, the former having early public relations and endorsement successes but the latter having a more convincing track record in the dirty business. Unlike calculating Ed, USPCI failed to effectively engage with African-American citizens who possessed less financial strength but were nevertheless the majority population. His outspoken supporter Brown capitalized on this, repeatedly condemning USPCI for not sincerely reaching out to his people. Yet Ed's Achilles' heel was the lack of verifiable assets and technical prowess to ensure that his team's proposal could rise above scrutiny in this contentious battle. True to his reputation for brilliant moves, the smart schemer from the Boro would not be outmaneuvered.

On March 22, 1991, Ed proudly announced that FTI had a new

teammate, a subsidiary of Hughes Aircraft Company. This spinoff brought immediate credibility to FTI's positive but cloudy scientific propaganda, boosting it with resources, technical know-how, and very deep financial pockets. Headed by executive James Abrahamson, a retired Air Force general, the new teammate knew that the topic was waste of the worst variety, and this leader tackled it head on, explaining the benevolence of the Hughes empire. "Providing environmental solutions is a significant part of our industry. I see this endeavor as an opportunity to benefit all mankind." What a marvelous gesture this seemed.

Ed's new collaboration became known as Hughes-FTI, and this unified consortium quickly announced that if they received the green light from the community and state agencies, they would establish the "Center for Environmental Optimization," promising that it would be a world-class research and development facility. Their new and improved concept might lift the distraught population in this smelly-water county into a better, crystal clear and refreshing

Choctaw protest march

future. This was a wonderful possibility as the establishment might develop fantastic chemical processes and brilliant storage solutions for a state and nation that would continually produce too much waste.

Yet regardless of all the delightful images of the economic prosperity that would follow the future dump trucks, be they driven by Hughes-FTI or USPCI, it seemed that this was still about hazardous garbage and money, not experimental exploration and discovery. This fear was evident when in the same month that USPCI arrived in town, a modest rally of opposing Choctaws reignited smoldering distrust for the waste undertakings, especially ones being proposed anywhere near lands sacred to their ancestors and sacred to them in the present.

CHAPTER 15
Promises and Science

Blackwell and others sought the support of everyone, including the underestimated Choctaw who had once ruled this part of Mississippi. Having rejected the National Waste Systems plan to use its solidarity, most members joined those opposed to any hazardous trash facility. Blackwell also dodged Brown and appealed directly to others in the African American population. Initially hesitant to oppose the earlier strong stance that Brown had obtained for FTI from the local chapter of the NAACP, black citizens quietly listened to Blackwell and the experts she and PEON were bringing in to counter the propaganda about new jobs and prosperity if any of the garbage facilities was approved.

It became clear that all of the opposition groups separated by class, wealth, and race needed to consolidate. They had to make this a distinct war, simplifying the battle to be between trash and no trash, incinerator and no incinerator, jobs and no jobs, and bad change versus the sanctity of their environment. Would the foul-smelling name of the county become a physical reality belching out disgusting aromas? That may have been too much for even the most neutral citizens in the area.

With the other waste entities, Ed saw that time would be on the side of PEON and citizens trying to stop the garbage proposals,

PROMISES AND SCIENCE

as with each passing month from late 1991 until the end of 1992, new revelations were discovered about the real science. It became especially clear that an incinerator and a dump for toxic stuff was likely more dangerous than anyone had anticipated. Earlier in March 1991, Blackwell and others recognized that the core issue was indeed environmental racism, and they enlisted the technical knowledge of national experts on waste management. Though most trash companies promised safeguards and insurance, it was explained that the risks outweighed the promised rewards, and if something went terribly wrong, any LLC trash company would itself become insolvent trash, its LLC insurance carrier would be overwhelmed and declare bankruptcy, and only the locals would deal with the bad, really bad garbage. No one on Earth could guarantee that an accident would never occur, and they reminded everyone of the frightening outcomes at Love Canal, Exxon Valdez, Bhopal Gas, Three Mile Island, and other sites of environmental disasters. It scared the wits out of everyone who paid attention.

Ed thought that his African-American supporters were totally on board, but that assumption received a surprising blow when a separate group of black citizens formed the African American Committee for Environmental Justice, and this group directly challenged the whole notion of black unity for any garbage facility. They passed their own resolution, comparing the endorsement of any waste endeavor to a modern-day burning cross, a flaming stick planted intentionally into the soil of a county full of disadvantaged blacks to scare them into acceptance of a perilous thing that would enrich not their own but white men in white shirts very far away.

PEON kept up its onslaught about the serious dangers of hazardous waste facilities. Perhaps the most damaging development for Ed and the other company leaders, yet the most encouraging moment for Blackwell and others with PEON, was the rally in county seat Macon on the perfect spring day of April 3, 1993. Held in front of the historic courthouse, the event became clear evidence of a united front rarely seen in Noxubee. The audience included whites, blacks, and Choctaw, though each race sat with their own. Participants also included protesting students from Indiana University who had learned about the land option Ed had arranged with their institution's Foundation. The rally was a rare social, racial, and county-wide collaboration of citizens. The next day was Sunday, and although politics were rarely addressed, especially in white congregations, numerous sermons in churches across the county that day made it clear that the Lord above was probably not in favor of any disposal concoction that would spoil His fine work in creating the area's prairies and hills. It must have been little fun for Ed and competing corporate heads when they learned of the success of the rally and the rising opposition from people of faith. Their schemes were now on hope-they-pass-away-soon prayer lists.

Church in Noxubee County

CHAPTER 16
Into The Trash Can

After three years of maneuvers with politicians and numerous influential citizens, and with the growing, authentic fears of possible long-term damage from this proposed industrial operation, vital regulatory approval had not been given to any of the garbage men. In April 1993, Ed announced that his consortium had not written the big check to purchase the gigantic property owned by the Indiana University Foundation. Activist students at IU had continually demanded that their fine institution should have no part in land acquisition for a dangerous facility. The embarrassed board allowed the option to expire, immediately sold the property to an individual in California, and then watched innocently from the sidelines as the new owner gave a refreshed purchase option to Hughes-FTI. But instead of a quick transaction, the process would move slowly, with one payment after another to keep the land purchase a possibility. Next, Ed's co-captain Hughes got a bloody nose when the national NAACP leaders accused the mammoth corporation of racial discrimination, and suddenly members in the local chapter rethought their earlier endorsement. Things were not going as planned for the Hughes-FTI alliance.

State politicians in Mississippi had seen the writing on the wall regarding any unwanted toxic waste facility, and they wisely knew to cool it on the scorching trashcan topic. It would be easier to drag their feet and delay an authorization to any of the competing garbage proposals. However, it would be a federal agency that would drive a temporary wooden stake into the corporate vampires interested in sucking the pristine purity out of Noxubee soil. From distant Washington, D.C., Environmental Protection Agency administrator Carol Browner announced on May 18, 1993, that all toxic waste permits were being placed in a box, one padlocked for the next eighteen months. This ensured that all dump trucks had to slow down, likely run out of fuel, and run off the road to crash beside the overjoyed clean waters of Dancing Rabbit Creek.

In late 1993, after spending a reported $23 million in engineering, lobbying, legal, and public relations expenses, Hughes-FTI took a painful pause. With the national moratorium, a delay of any hazardous waste facility had been achieved. One would expect a celebration in Noxubee County, yet no one really won. The whole episode had opened deep past and contemporary wounds about land ownership, screwed Native Americans, never-compensated African Americans, three-pronged racial inequalities, supposed corporate benevolence, and the reality of corporate greed. The deep cuts might eventually heal, but the scars would need curative creams from the best white, black, or Choctaw medicine man.

The EPA moratorium was not permanent. Similar to the one Mississippi had issued in the mid-1980s, this national delay would expire, this time in just over 500 days. As 1993 ended, Ed departed the scene; this was followed by a surprising peace treaty when Hughes-FTI shook hands with USPCI. Now a merged enterprise to deal with toxic waste, this corporate creature would quietly wait like a single, fat vulture on a power line in rural Noxubee County for eighteen months to pass.

Whether Ed's departure was voluntary or not, by leaving the area he avoided the hatred of an immense number of Noxubee citizens. He seemed to possess an almost magical Kevlar-bulletproof ability to withstand the anger of opponents, be they individuals or organizations. If any of the waste entrepreneurs had succeeded, the fury would have only intensified. Yet that had been avoided for the moment, and most grateful for the outcome in this county would be spirits who would always hover above this territory, joined by the souls of long-dead whites and especially enslaved blacks who had also toiled to transform this rugged landscape into a place to grow and live, not as a dirty spot to burn and dump. (Years later, a waste incinerator was established in Mississippi, but the land of stinking water would be spared.)

The failed waste endeavors in Tennessee and Mississippi left long trails of unfulfilled agreements, furious losing investors, and broken county relationships. I had never heard of Ed's exploits in the counties of Giles and Noxubee until I read his obituary in 2014. In recapping the life of the murdered classmate, it mentioned that smart Ed was the main subject of a book published almost two decades earlier in 1996, a 410-page publication entitled *Uproar at Dancing Rabbit Creek*. It was written by environmental law professor Colin Crawford, and it received

Colin Crawford's book   AWP

glowing reviews for its coverage of racial tensions, economic disparities, and undeterred courage. Although I studied newspaper articles and had many interviews with individuals involved in Ed's toxic chases, I want to acknowledge that Crawford's documentation provided the best and most thorough information. His superb book also had striking similarities to another environmental battle, one that became the basis of the Academy Award-winning film of 2000, *Erin Brockovich*.

As Ed headed north back to Tennessee, a completely separate, enormous garbage landfill would eventually rise in his hometown. He would not be a participant in the growing eyesore, but its success would ironically be fertilized by a fellow graduate from Central High's last class back in 1972. Could it be that both of them had learned not only trash talk but also trash actions while in high school?

CHAPTER 17
Positive Fallout

With a dumbfounding attitude of courage and fortitude, Ed was quite pleased with himself and proud of his bold yet failed endeavors in both his Volunteer State and the adjacent Magnolia State. He had almost accomplished something that was truly needed, a place for bad stuff to be handled and safely deposited. He was never shy and felt he should be honored for marching right into battle to bravely establish a facility that would always be opposed anywhere and everywhere by environmentalists and not-in-my-backyard "NIMBYs." Ed had been in it to win it, and he'd given it his best shot. It didn't hurt that he had made the attempt almost entirely with others' financing. Even with huge bullet holes on the FTI balance sheets, Ed presented copies of Crawford's book—which documented in truthful yet quite unflattering ways the numerous tactics and the expensive collapse of the two projects—to his friends and business associates.

Ed closed his Nashville office in 1994, and my design firm also moved the following year from that convenient midtown area into the historic Elm Street Methodist Church, a large structure erected in 1860 and ripe for a resuscitation by crazed designer types. From Ed's new location in the Boro on the uppermost floor of a commercial office building on East Main Street, he made frequent journeys around the nearby public square and at eateries throughout the city and surrounding county to make sure he was seen and available. He was out and about, seemingly without doubt. Advertisements in the local paper spoke of the man's unquestioned knowledge in his specialized trade, and in 1995, he was named "Agent of the Year" by the Guardian Insurance Company. Ed was back in his saddle.

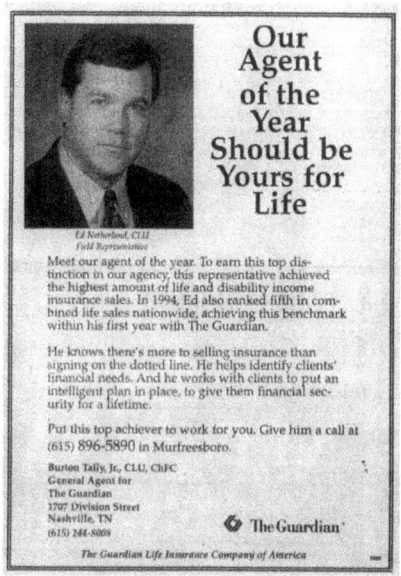

"Agent of the Year," 1995   RCH

Even with the investment disappointments of the hazardous waste projects, Ed had succeeded in finances and health. He was a determined fighter and had battled melanoma with the valuable help of Vanderbilt Medical Center oncologist Dr. Robert Oldham. The doctor's recommendation of a new vaccine developed by Duke University was a gift from above, and Ed experienced a full recovery, quite rare at the time. Of the numerous participants in a select group to receive the experimental treatment, he was the sole survivor. It may have been his finest moment.

With the storm of the disease blown aside, Ed's energy and mojo returned. His ability to form new business alliances and obtain new funding sources would later appear in the annals of his profession. Though he continued to think about the incinerator idea, he wisely dropped that expensive girlfriend and refocused on his original path to prosperity. Constantly aware of ideas floating across the nation concerning life insurance, he learned of an interesting new one in the clouds that had been used by several huge companies. As with other insurance products, the "pro in the Boro" gained a full knowledge of the process and introduced new clients and especially larger companies, educational institutions, and elected governmental officials to this creative mist that had the potential to douse raindrops of gold. It was a showering from above of money from these policies, and Ed dove into preparing proposals to insure that not one but everyone was signed up, filling to the brim a gigantic lake teeming with thousands and thousands of premium-producing policies. That deep pool would be full of creatures called COLIs.

CHAPTER 18
Dead Janitor Insurance

Back in the 1980s and 1990s, large corporations including Walmart had implemented a process in which they paid very reasonable premiums for life insurance policies on all of their 350,000 employees. Most of the workers were young and receiving minimum wage, and some were part-time workers. The company offered a free $5,000 life insurance policy to employees who agreed to the coverage, and only a few hundred did not sign up. It seemed like a very benevolent act for surviving spouses and children. Most businesses, including my architectural firm, possessed "key man" term life insurance policies on the partners and on the most important contributors to the success of the company. Upon death, the business would receive a big check to recover and find a replacement leader, and the surviving spouse would also receive a large amount. With that payment, he or she, albeit bereaved, would be financially compensated and, most importantly, out of the picture. This much larger scheme for coverage of not just vital but *all* employees was called "corporate-owned life insurance," commonly abbreviated by the acronym COLI. It seemed rational and fair until one looked behind the corporate curtain and under the sheets covered in small print. What Walmart, CM Holdings, Trans World, Dow Chemical, Winn Dixie, Nestle USA, Procter &

DEAD JANITOR INSURANCE

Gamble, and an auditorium full of other corporations may not have fully disclosed to employees was that the business entity would not receive the same amount paid to the survivors but, instead, might receive a far greater amount of $50,000 to upwards of $400,000. This payout determined by the age of the deceased employee and actuarial charts that the average Joe, like me, would never completely understand.

The scrumptious payout flowing from this mainly unknown, tilted arrangement finally got bitter when the family of Felipe Tillman, an employee of Camelot Music store in Tulsa, Oklahoma, received almost nothing when he died in 1992 at the age of 27 of complications from AIDS, yet CM Holdings, the parent company that owned Camelot Music, received a jaw-dropping $339,000 because it had a death policy on dead Felipe. These lucrative policies, negatively called "dead janitor insurance," were offered by Hartford Life, Lloyds of London, Travelers, and many other similarly strong and stable insurers, and some of the corporate purchasers validated the policies and the wildly unbalanced payouts by placing the big bucks from deaths into employee retirement funds or other "nice" programs. Outraged, Felipe's relatives slung multiple lawsuits at the Goliath companies. The best rocks in their slings were engraved with those most terrifying initials: IRS. Lucrative write-offs allowed by the federal government had initially made the dead janitor policies attractive, yet regulators learned of the many odd and inappropriate uses of the payoffs.

Due to scathing articles, including one by Ellen Schultz and Theo Francis in the April 19, 2002, issue of *The Wall Street Journal*—most certainly perused by well-read Ed—these seemingly greedy policies became poison ivy in public and employee relations. Wise corporate attorneys quickly cut the vines, and most COLI policies of this tilted nature became something to be avoided or handled only with thick gloves. However, the technique wasn't totally illegal, and a different variety with better financial and improved moral clarity seemed fine for organizations struggling to deal with employee benefits, rising medical expenses, and pensions. To address those national challenges, and knowledgeable of the pros and cons of large COLI creations, Ed was ready to share a more balanced arrangement and convince entities to sign up. The premiums for these policies only added to the fortune he was amassing, and with even more confidence and contacts, the sky was the limit.

Text on COLI

CHAPTER 19
Reunion Justice

The last time I saw Ed was many years after his Mississippi adventure at our twenty-fifth high school reunion. The event was held in the early summer of 1997 at that same country club where, three decades earlier, four teenage boys had kept the surrounding fairways nice and tidy. The year before, I was part of the dedication ceremony for a new, large urban park that I had designed for underappreciated property in Nashville directly north of the landmark Tennessee State Capitol, and a few in attendance knew of this and offered nice congratulations. Ed arrived fashionably late and characteristically parked his convertible Mercedes near the front door in the restricted fire lane. He wandered through the reunion, flirting—as many of us did—with the charming and still-lovely former Central High coeds. He soon vanished from the party. I and most others were completely unaware of the events in Noxubee County, and as far we knew, forever proud yet haughty Ed

Central High 25th Booklet

was a very successful insurance businessman with an expensive car, big house, sprawling farm in the nearby rural Lascassas community, and a validated large ego.

Central High 25th reunion   RB

I never saw Ed at the numerous following reunions every five or ten years, ones that continued the theme of "WHERE WERE YOU IN 1972?" All of these festive and enjoyable Baby Boomer events were reinforced by the justice of passing time, effectively separating the wheat of accomplished, warm, and successful classmates from the chaff of jerks, snobs, thugs, cheats, and losers. Although those past days at Central had a decent level of racial harmony, this reunion and the others would sadly be mainly Caucasian occasions. Regardless, the parties were highlighted by the special appearance of graduates who had become famous locally and far away. Ed's local notoriety would also increase as would another classmate with a similar name, and this other one would receive fame of a dirty kind due to a refuse disposal challenge much like the one that the first Ed had tried twice to capitalize upon.

CHAPTER 20
The Other Ed

Surely it is only a coincidence that another Ed in Murfreesboro, this one with property near the rural community of Walter Hill, would make a deal with a waste-dumping business. Epps Edwin Matthews III was our fellow classmate at Central High, and he and his brother inherited property on East Jefferson Pike just off US 231, the highway connecting the Boro with the city of Lebanon only 32 miles directly north. Next to their sizable parcel in the community of Walter Hill was a slowly growing behemoth, becoming almost beyond imagination: a regional landfill, a colossal dump.

Eddie Matthews, CHS 1972

In 1989, Browning Ferris Industries bought land in Rutherford County and established a much-needed landfill that became the single major dump for one-third of Tennessee, a jaw-dropping 39 counties in Middle Tennessee, including the most significant contributor, Metropolitan Nashville & Davidson County. A national corporation, BFI had 25,000 employees and 90 other landfills, so they knew all aspects of the business of garbage. No one could miss their extremely visible and constantly growing mountain of garbage

NETHER LAND

BFI trash truck

called the Middle Point Landfill, prominently located right beside US 231, what locals called the "Leb-non" highway. Ed had dealings with BFI, but he wasn't involved in this trashy establishment. His hometown, the Boro, was very proud of its authentic standing as the geographical middle point of Tennessee, but the reality of this place being another "center" was definitely not what Ed, the Chamber of Commerce, the Visitors Bureau, and almost every citizen of Rutherford County had in mind.

The slowly growing monster of refuse was only a short distance north from one of the county's signature establishments, the U.S. Veterans Administration's famous Alvin C. York Medical Center. Since 1940, this important facility has served soldiers, sailors, and airmen on its pastoral, 321-acre campus, and as an attractive architectural treasure was listed in the National Register of Historic Places. The distinguished brick-clad structures at this special place for those who served

Geographical center of Tennessee

their nation also looked good, and an honorable pride of service was always in the surrounding air. Sadly, the stench of decaying garbage from the neighboring BFI dump was also occasionally in the air, not the most healing atmospheric medicine for the nation's finest.

Alvin C. York Veterans Hospital

Middle Point Landfill became a visually overwhelming, smelly manmade mountain, receiving daily trash brought to the place by the subsidiary's enormous fleet of blue trash trucks, each stamped in white with the giant initials of BFI. The landfill had its own website noting environmental stewardship, happy employees, support of good causes, and civic participation, yet the odorous place was a featureless brown landform that rose more than 200 feet high and startled anyone traveling past it for the first time on the busy highway. "Mount Trashmore," as disgruntled residents tagged it, would never be featured as a local landmark on promotions about the beauty of Murfreesboro and Rutherford County. The other Ed would help it grow even larger.

NETHER LAND

"Mount Trashmore"

On April 4, 2000, our former classmate Eddie sold his adjacent 344 acres on East Jefferson Pike to BFI for the pleasant sum of $13 million. This increased the total land owned by BFI to 1,147 acres. His land was bordered on its south edge by the East Fork of Stones River, the same waterway that, upstream, defined a similar boundary of Ed's picturesque farm in nearby Lascassas. Many former Central classmates were furious over Eddie's cashing in, and several times I heard that the ginormous landform should now be known as "Eddie's Dump," not the lasting heritage anyone would desire. But the now quite wealthy classmate apparently couldn't care less, cashed the fat check, got into a lawsuit with his brother, perhaps got into other "harmful" substances, and then nine years later on March 25, 2009, vanished. Eddie had moved to Bedford County, and his ATV was found along the banks of the nearby Duck River. It may have been an accident, but foul play wasn't ruled out as his body was never found. Our Ed was lucky to have property upstream from the enlarged landfill; eventually everyone downstream became outraged when testing of this fork of Stones River's water revealed traces of

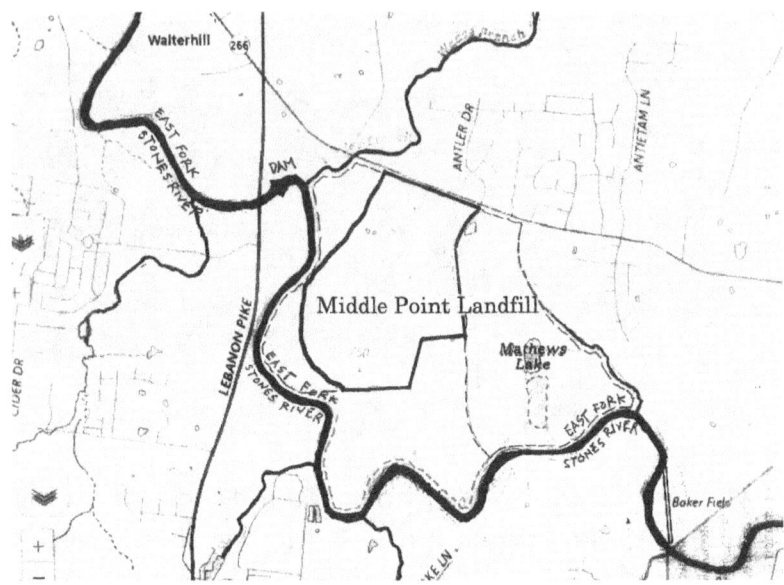

Stones River near massive landfill   *TSLA*

low-level radioactive chemicals. Due to this pollution, long-term environmental worries, and horror over this gigantic mountain of rubbish that had emerged in this otherwise pleasant, soothing county, citizens rose up. BFI was denied permission to expand its landfill in September 2023 by the Regional Planning Board. The landfill would be at capacity in 2027, and without any other soil to spoil, it would close, with hopes that Mother Nature would do her best to eventually reclaim the mountain of trash. Meanwhile, Eddie gone missing might not ever be missed, his face not on postcards but perhaps with missing kids on milk cartons.

CHAPTER 21
Lilac Flowers

The main Ed was back at it in the early 2000s. His reputation in Rutherford County was solid, and he was asked to join an ad hoc governmental committee in the evaluation of its financially troubled nursing home for indigent residents. The sale of COLI policies had been quite profitable, yet inventive Ed was always planning for more. After research and consultation with his legal and accounting experts, he formulated financial instruments that would be coined "LILACs," an acronym for life insurance and life annuities certificates. It was a very complicated and initially hard-to-understand concept combining insurance policies and premiums with investment annuities. The LILAC technique promptly caught the attention of the insurance industry and reporters who covered such financial offerings.

On June 5, 2004, reporter Stephanie Strom of *The New York Times* composed an overview of the new concept, crediting Ed for his creativity and leadership with this bold yet under-tested life insurance product. The musician Prince had made the phrase famous in his 1984 album, yet the smart

LILAC
Life Insurance Life Annuity Certificate
The holder of this policy will receive a return on a forgoing any previous agreements, the limits of th ongoing adjustments as determined by the Consui

LILAC policy

fellow from the Boro had invented an equivalent purple rain, one that would shower premium payments into his bank account. The technique also assisted charities, and due to Ed's urging, officials at Donelson Christian Academy, a private school in Nashville, were contacting donors to purchase policies whereby the death proceeds would go directly to the school. Strom noted that five LILAC deals had been completed in 2003, and a sixth deal was being planned for 2004. Prosperous alumni of the University of Texas, known as "Texas Exes," were signing up for Ed's purple insurance flower, and the group expected to raise $80 million over the life of the contracts, a home run that would greatly increase the university's endowment. The alumni organization noted their appreciation of Ed's product, happily noting that just three participants in LILAC transactions might alone generate as much as $1.8 million. However, an influential trade publication, *The Insurance Forum*, voiced serious reservations over the long-term implications of such policies. Two national insurance associations also issued their opposition to the whole thing due to questions of financial transparency.

Purple Lilac flowers   PEX

Never shy of controversy, and confident of the returns that would sprout from his invention, Ed carried forth with this legal and very profitable LILAC creation. Due to this acronym, I was immediately reminded of a distinctive lavender corsage I had pinned on a girl long ago. Now the mention of Lilacs reminded me of the power of the color purple. The floral analogy might be a stretch, yet this hue has remarkable relationships, meanings, and associations with success and royalty. It is what gives violet, heliotrope, hibiscus, hyacinth, and other blooms their remarkable distinction as a flower more regal than any other. Pollination of his unique floral species proved quite sweet and tasty for Ed.

CHAPTER 22
Blooming Fortunes

In 2004 and 2005, Ed closed five deals using the LILAC instrument, resulting in insurance sales of a reported eye-popping $1 billion. Commissions from the premiums arrived on a regular timetable. For him, life must have seemed wonderful. In 2006, he opened a new Nashville office in the Palmer Plaza building at 1801 West End Avenue only two blocks from his earlier location on Church Street. He continued to present his expertise and his Certified Life Underwriter credentials, and this new address was also listed in the phone book for InsCap LLC, one of his limited liability corporations. He was diving into the use of another controversial method for investors to join the investment arm of the life insurance world. He and others promoted "viatical settlement investments," an unpleasant but sometimes urgent escape hatch for often terminally ill individuals whereby a third party would pay the sick person—often wanting the cash or in desperate need of it—a discounted lump sum to purchase the life insurance policy;

Bentley convertible

later when the ill person perished, the purchaser would collect the full death payment. Ed's insurance creations and other investment strategies were yielding bushels of wealth. Locals back in his home in the Boro noticed his fortune, and ever the showman, Ed cruised in a six-figure Bentley convertible, likely the first ever in Rutherford County. He made himself available to investors across the country, and many lined up to hop into the exquisite vehicle and pay the steep admission fee to pick purple flowers from the chauffer's flourishing gardens of insurance devices and investment opportunities.

1801 West End Avenue

Ed had traveled extensively in his dealings, many times on a private jet, to meet those who possessed tall stacks of money and needed a place for the precious dollars to silently and legally rest and

230 Park Avenue   *WIK-P*   *Friars Club*   *WIK*

multiply. He made numerous connections throughout the nation, and he had favorite places to wine and dine those who expected to be wined and dined. He opened an office in the prestigious New York City landmark Helmsley Building at 230 Park Avenue, had an apartment on the top floor of a nearby hotel, and schmoozed with the rich and influential at the elite Friars Club. Ed sought the fattest of fat cats to share his capabilities and expertise in lucrative strategies to help them make more money and keep as much of it as possible away from that greedy, conniving, and hated relative, Uncle Sam.

In 2005, Nina Investments LLC, headquartered in New York City, was persuaded to invest in one of Ed's offerings, an insurance premium-financing product named "Ultra." The LILACs and other techniques had apparently worked well, and so should this new item, another derivative that provided permanent protection with equity-linked flexibility and attractive tax advantages for the very wealthy. Ed had been joined by another brilliant and charismatic fellow, Ira Brody, a longtime associate of former New York mayor Rudy Giuliani. This impressive individual had moved from the Empire State to the Boro to provide additional management and leadership in the growing orchard of profitable companies that Ed and his other colleagues were cultivating. Brody also wanted to jump into the political arena, and with valuable experience under the Big Apple's big man, he developed ties with Republicans throughout Middle Tennessee. Ed and Brody were on a roll with the new Ultra sprig and a bouquet of similar insurance, investment, and financing species.

According to the *Chicago Sun Times*, Nina invested $6.5 million into this promising insurance-based scheme; this was followed in

2006 with a much larger contribution of $75 million, with promises from Ed's team of hefty returns in only a few years. Based on the success of the Nina endeavor, Ed's company Lilac Capital LLC would morph, sprouting new hybrids including InsCap Management LLC, ISM Advisors LLC, and several others. These limited-liability insurance and investment entities were changing and rearranging in legal, financial, and accounting ways that would create an intertwined and profitable thicket, one thorny enough to give the uninitiated person like me a piercing headache.

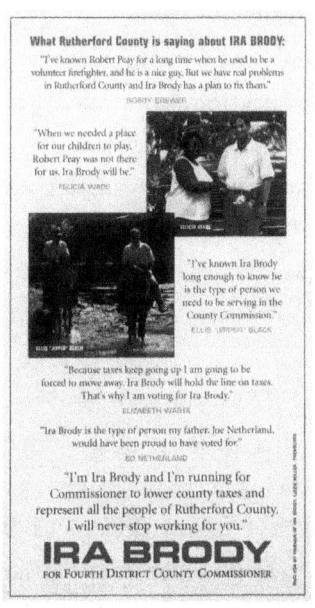

Political ad for Ira Brody

The variety of investment opportunities offered by Ed and his associates seemed to be completely legitimate. They knew that although convoluted, these multiple corporations provided an extremely smart way to guard one's flanks while keeping the chess board of business and investing both interesting and promising. Of course, with any corporate plant tagged with the LLC appendage, "limited liability" meant that there was only so much money available to investors if financial managers over-clipped the numbers or if the Internal Revenue Service's special formula of Roundup investigation was sprayed and the entity of invested green dollars withered and dried up.

In April 2009, Ed and InsCap's Harish Raghavan filed for a patent on their "systems and methods for enhancing group benefit plans" using life insurance funding. This new financial sprout, though unnamed, was in the same vein of his earlier creations, and with patent protection, continued success seemed inevitable. Coincidentally, smart Raghavan had been a graduate student at Penn in 1981, the same year I was there obtaining the Master of Architecture degree. He was in the Master of Business Administration curriculum, and we may also have been in the same large Wharton School class on real estate, one to which I had been asked to attend and provide a non-business "designer" perspective during classroom discussions.

With the new, still-unnamed technique and existing profitable Ultra product both under control in their InsCap entity, Ed and Raghaven were enjoying their ingenuity, slowly waving a wand above an overflowing caldron of liquid cash. In 2010, Raghavan left InsCap to become the founder of Gracie Point Holdings, another company offering insurance financing. For unknown but probably smart strategic reasons, InsCap was fragmented, changed, or added to with an entity called Concord Capital Management. Yet from their seemingly indestructible castle of prosperity, Ed and his partners and associates with Concord may not have noticed the monetary drought headed straight to Wall Street, one that would eventually dry up or poison investments of almost all varieties and species, including theirs.

CHAPTER 23
Come Tumbling Down

With the arrival and fallout of the financial fiasco that roared during 2008, 2009, and the next few years, over $80 million of Nina Investments' funds in the Ultra product vanished into smoky air. The plum delicious creation plummeted into the fire, and its aroma of investment success was replaced with a most unpleasant, foul odor bellowing from an incinerator of greenbacks. The Great Recession might have been the culprit in the enormous financial loss, but angry Nina officials were not convinced. Although Raghavan may have not been involved, Ed and business associates Ira Brody and Matthew Ross became entangled in a series of lawsuits in 2010 and 2012, legal eruptions with widespread shrapnel and finger-pointing over how and why the huge investments had gone up the chimney. Litigators accused the three of squandering the money for personal gains and cushy lifestyles, and their collapsing castle now seemed to have been made of sand. Because Matthew Ross was employed by Fifth Third Bank, he became a most obvious target of lawsuits, with allegations that his bank had not properly monitored lines of credit to his amigos, expanding the complexity and damage of sharp barbs in the already bloody courtroom fights. Naturally, Fifth Third Bank sent in lawyers who together created the equivalent of one angry queen bee, responding that their financial organization should not be responsible for those who illegally obtained honey, and that Ross was a bad, lone drone and not part of the trusted hive of bank officials and employees. A few settlements would occur over the lost honey, but some never would, and Ross reportedly may have left the United States. To my knowledge, there were never criminal accusations against Ed and his cohorts, and it may have been that they had simply enjoyed the fruits of the investment arrangements, outmaneuvering and legally enjoying the money honey. If so, would congratulations be in order for their slippery triumphs?

> **Lawsuits by Nina Investments**
> Significant legal action has been filed by Nina ⟋ Murfreesboro financier Ira Brody and Fifth Ban Ross for losses of approximately $80 million. E Murfreesboro insurance executive Ed Netherl: using Nina's investments in a premium financi ... TDA" for personal uses and la*
>
> *Newspaper coverage*

Ed eventually closed his Nashville office, somehow surviving the grueling financial and legal onslaughts, and resurfaced to be at it once again. From discussions with professionals in the Boro including real estate agents, attorneys, scattered public officials, and others, I learned that Ed had reestablished a decent reputation. Now he was gallivanting on a few local endeavors, but most of them related to his expertise in the insurance trade. As always, he exuded a classy, brash personality, and he was always appropriately dressed, displaying a captivating cloud of knowledge and confidence. I don't know if he sent gold to local charities, but according to many, Ed always liked to pay for entertaining others, achieving a notoriety and appreciation for his graciousness, big smile, big handshake, and supposedly big bank account.

Since Ed and I had existed in different universes, I knew almost nothing of his personal life, kids, close friends, or hobbies. However I discovered some insight about his personality and ingenious financial capabilities from certified public accountant Ralph Turley. My father had depended on Ralph for his tax preparations, and after he died in 2008 and my mother passed away only eight months later, I spent many hours in this CPA's office. It was located on Lytle Street, the same road that, a few miles away, bordered the east side of Campus School and the north side of Central High. Ralph helped me work through the various issues of inheritance, and I was still meeting with him on final complexities in January 2015, only two months after Ed's mysterious death. Asking if he ever knew the murdered insurance tycoon, Ralph shared that they had a good business relationship, and he had served as the trustee of Ed's big farm in Lascassas. He had also invested small amounts into some of Ed's financial adventures. Of course, he never shared confidential information, but I had the impression that prior to Ed's demise in late 2014, my former classmate may have experienced serious financial challenges.

Regarding benevolence and charismatic reputations, Turley and I also recalled investor Bob McLean, another person in the Boro who, like Ed, enjoyed being beloved as a top gun. Ed must have known about this other generous soul in town because hot shots tend to gather in a league where the members want to be on the very top of the pole, whatever the pole is. The desire for respect and admiration shaped both big dudes and their engaging mannerisms and physical presences that couldn't be missed. Ed was a grizzly bear and McLean a big teddy. To my knowledge, Ed managed his own investments and escaped the other animal's tomfoolery. Others, unfortunately, did not. McLean's story is also one that would involve my own family.

CHAPTER 24

The Need Greed

Among the great motivators that affect everyone are fear and greed. Ambition can get out of control, and fear can make good people commit bad deeds. Ed was definitely ambitious, but he had always seemed to stay within legal guardrails. Not so with Bob McLean, one who had a special, unsatisfied greed. Indeed, for some, greed is for the need, a need to be appreciated and beloved.

A friendly yet mysterious individual raised in nearby Shelbyville, McLean established in the 1980s a small investment office in the Boro. He had attended MTSU, briefly played in a small folk band, and had been a member of the fraternity with the celebrated "Old South Ball" which had lured away my hometown honey years earlier. McLean's fraternity brothers became prospective investors to which he offered special, high-yield promissory notes. Soon the investments produced returns of fifteen to twenty percent, and the extremely pleased clients urged friends and associates to also board McLean's financial airplane and soar into the puffy clouds of dizzying high-altitude paybacks.

Philanthropist Bob McLean   IMG

When the twenty-first century arrived, McLean started giving significant donations to charities and institutions. In 2002, he presented $1.5 million to the noted music department at MTSU, allowing his alma mater to buy 54 Steinway pianos. He also started supporting needy music students with tuition and living expenses. His yearn was to be yearned for. McLean's most publicized gift was in 2004 when he presented a $1 million check to the Country Music Hall of Fame and Museum, the still-new 2001 building designed by our architectural studio under lead designer Seab Tuck. With that check, the museum purchased "Mother" Maybelle Carter's 1928 Gibson guitar, Bill Monroe's 1923 Gibson mandolin, and other one-of-a-kind instruments. The formal announcement on August 23, 2004, was truly joyful, a celebration to thank the rich teddy bear for allowing these rare musical items to be purchased and added for display in the most beloved institution devoted to America's distinctive homegrown music. After the donation ceremony and national media coverage on that special day, CMHF board members were still elated, and I can imagine that they happily lost consciousness, falling onto the end-gain wood floor of the museum's most hallowed chamber, its rotunda Hall of Fame. Renditions of the unique museum and its new priceless gifts owned by originals Carter and Monroe soon made for popular takeaway gifts and postcards.

Country Music Hall of Fame   ADG

THE NEED GREED

McLean became a "social butterfly" and appeared at most area galas, fundraisers, and political events in the Boro. He convinced grateful MTSU officials in late 2005 to name the recital hall in the Wright Music Building in honor of one of his favorite music professors, T. Earl Hinton, my father. McLean had never offered his splendid investments to my dad, and our entire family was overjoyed and humbled when Hinton Hall was dedicated. So many of us had affection for this jolly, gregarious philanthropist. He continued giving wonderful gifts and helping finance worthy undertakings. He became the Boro's Santa Claus. Unfortunately, the need for prestige, notoriety, and admiration had a very sad, dark side.

In early 2007, investors learned that McLean was running a Ponzi scheme. His house of cards collapsed as it was revealed that he had spent colossal amounts, constantly moving funds around and issuing fraudulent statements to his victims. His issuance of unsecured promissory notes to build his empire had allowed him to avoid regulatory oversight. Now the only oversight was by the outraged investors who had trusted McLean as they looked over his shoulder onto the blank sheets of financial reality, horrified to have been so easily tricked and cheated by this rotund, fine fellow now revealed as an extraordinary con artist. McLean had smashed the financial nest eggs of more than fifty victims to the tune of a whopping $67 million. It was over, and he was over. With egg on his face and even larger lawsuit omelets on the way from damaged and furious individuals, McLean exited the Boro on September 25, 2007, drove south 26 miles to his hometown, parking in the rear of Shelbyville's First Christian Church, and ended his life using a 38-caliber pistol.

The tragedy of the gut-wrenching legal and financial mess

afterwards was of the worst kind, and an article about the disaster appeared in *The New York Times* on November 10, 2007, noting the prestige and power of those who said they had deep pockets and the calamity that befell everyone when those pockets were found to be completely empty. When Christmas that year arrived, the real Santa would fly over Middle Tennessee while the sad imposter's body and reputation would lay in a cold box six feet under. And only one year later near the end of 2008, Bernie Madoff was arrested in New York as the mastermind of the largest known Ponzi scheme in history, swindling an estimated $65 billion through similar techniques used by the local one, "Murfreesboro's Madoff."

The dead teddy bear McLean had possessed a cheerful, "gosh darn" reaction to his lavish gifts now known as fraudulent. Ed also enjoyed being generous. His past and future actions may have resembled some of McLean's salesmanship smoke-and-mirrors techniques, yet Ed was too smart to be so dumb and to do something so blatantly illegal. More outgoing and boastful, he always delighted others when as the big dog, he picked up the restaurant or bar tab. McLean was the very definition of braggadocio, showing off a wealth that wasn't there. Ed was instead focused on legally making tons of real moolah to be deposited in real accounts. With that, he might do similar beneficial deeds that had allowed the sun to shine brightly on McLean. How fun it would be for any of us to be Saint Nicholas.

CHAPTER 25
Tennessee Proposal

Ed still had many valuable political connections with Tennessee officials who regulated the sprawling insurance industry, an ever-changing business that had its heroes and villains, honorable ones and crooked shysters. Much earlier in 2003, he had planted an idea by persuading officials at the State Capitol to hear his proposal. It was to offer a "new and revised" version of the once-tainted corporate-owned life insurance technique to the state to cover all of its employees. This COLI scheme for a government was not unlike other comparable policies that had been purchased by corporations and institutions, but it had a significant difference as potential medicine, a sugar-coated tablet for what always gives state official tummy aches: the allocations for government employee pension funds. Smart Ed's scheme involved the issuance of secure Tennessee state bonds that would finance present and future premiums, and the life policies on all state employees would pay an equivalent amount to the deceased worker's family and to the state. This new *government-owned* life insurance strategy could be known as GOLI, strangely reminding folks of what the amazed, simpleton Gomer Pyle would often utter on *The Andy Griffith Show*. What an incredible home run the program might be, as Ed would be guaranteed a slice of the premium payments almost forever.

TENNESSEE PROPOSAL

No action was initially taken as the proposal would need a change in state insurance regulations to be lawful. Surely that day would come, and Ed politely and continuously suggested the idea to those in power at the State Capitol.

The crowning moment of this possibility arrived in early 2009 when the frequently-pitched GOLI scheme became part of a political contest in the historic tall halls of architect William Strickland's masterpiece. It involved Ed's proposal—originally under his Lilac Capital LLC but now under the entity of InsCap Insurance Services—and it involved his bud, Ira Brody. The concept had its good points, but due to Brody's politicking to become the new state treasurer, opponents sliced and diced not only the concept but him. The GOLI pill wasn't swallowed and instead hawked into one of the ancient, tarnished, and nasty brass spittoons in the statehouse. In a mixed government of blue dog Democrats and deep red Republicans, the idea received a unified stamp of rejection signed with rarely-seen purple ink. With his political ambitions in Tennessee derailed, Brody eventually returned to New York, far away from Ed and his forever-evolving insurance and investment concoctions.

Grand hall of State Capitol   WIK

CHAPTER 26
Island of Paradise

At some point in the midst of his wheeling and dealing, Ed became enthralled with the paradise found in the U.S. Virgin Islands. Having traveled all over, he had enjoyed some of the most breathtaking scenery the nation had to offer, and man of taste that he was, these spectacular territories were apparently among his favorites. Many inhabitants and visitors at these luxurious island playgrounds of St. Thomas, St. John, and St. Croix were from the

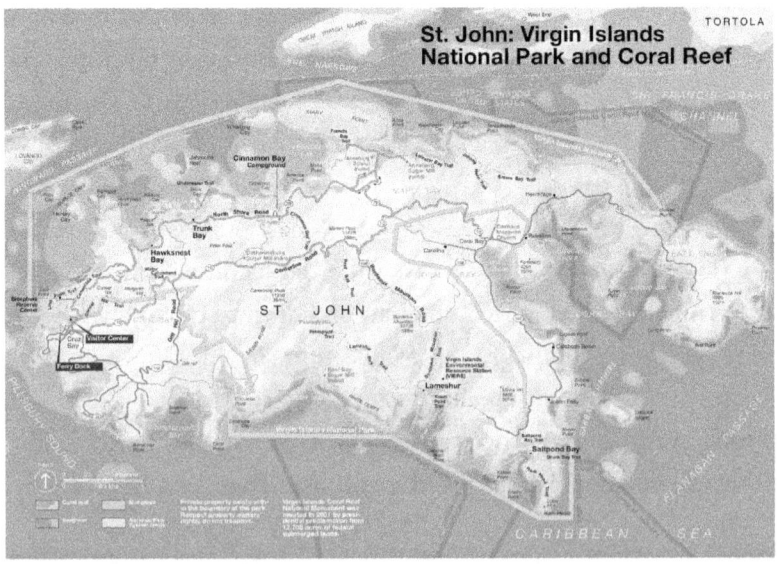

St. John Island   *WIK*

uppermost crust of society, the rich and the famous, the Who's Who and the ones who could do. Similar to the Friars Club in the Big Apple, here Ed could share business deals and investment opportunities in a discreet way to deep-pocket prospects having enormous fortunes on the main islands or other nearby ones. I have no knowledge that he ever exchanged his ideas with the most prominent residents, but it's an impressive list, including entrepreneur Richard Branson, actor Kelsey Grammer, model Hannah Jeter, musician Kenny Chesney, and perhaps the most influential tycoon possessing valuable worldwide connections from his nearby private resort island of Little St. James, the soon-to-become-notorious Jeffrey Epstein.

Island paradise ZAZ

Ed was obviously captivated by St. John, the smallest of the three main isles and located directly east of big brother St. Thomas. True to form, Ed glanced around and saw yet another way to strike it rich, this time an adventure to create a wonderful and potentially quite profitable residential development. To plant authentic and legitimate roots in the soil of this "off-shore" pasture, Ed established in January 2013 the foreign for-profit corporation of Sea Mist V.I. Inc. from a stateside, luxury residence in Jacksonville, Florida. Curiously, this new entity had an address and a female corporate treasurer found far away in a United Parcel Service mailbox in Franklin, one of Tennessee's most prosperous towns, located only thirty miles directly west of the hometown that Ed and I shared.

Soon, Ed also established East End Limited, LLC, a real estate company with its official address listed at his farm in Lascassas. He and New Jersey development partner Mike Davies had earlier purchased many expensive acres on the most remote portion of the nine-mile-long paradise landmass of St John. This narrow finger of utopia was known appropriately as the "East End," where sunlight would first illuminate the treasured island. It was a gamble in that most of the homes and attractions on the nineteen square miles of St. John were on the west and south parts, and sixty percent of the isle had been protected in 1956 to become the Virgin Island National Park. To make the deal for this most "secluded" development on St. John work, Ed needed to tap into the territory's economic incentive program, one requiring annual residency for a minimum of six months. With enough financial watering, this controversial development—opposed by many on St. John—might have germinated. For unknown reasons, it did not.

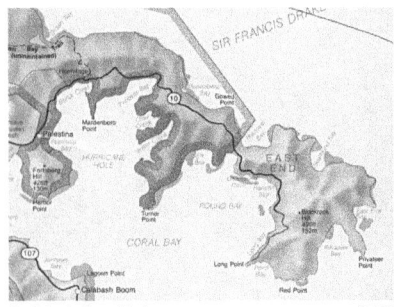

St. John's East End   ZAZ

The Virgin Islands possess a fascinating and dark past, one that Ed may not have studied or fully appreciated. Similar to Ed being a tad unaware of the contentious history of Noxubee County, here it was déjà vu all over again. Christopher Columbus sailed past—and named—this lovely archipelago on his second trip to the New World in 1493. Others from Europe would soon visit. With no ruling authority recognized by arriving Europeans, and with many places to drop anchor and hide, the calm waters would be visited by numerous scoundrels during the late-1600's "Golden Age of Pirates." Danish planters occupied St. John in the 1700s, enslaved the natives, and with imported slaves from Africa, established enormous profitable sugar plantations.

Though slavery was abolished in 1848, myths remained of supernatural forces, evil spirits of malevolent entities, and slave-masters. A superstition spread that such bad forces and thieves might be discouraged by many varieties of deterrents including the placement of an apotropaic, blossom-like hexafoil symbol on the front wall of one's house, or by the hanging of a strong-smelling wreath containing garlic, camphor, or native sea lavender over one's front door. Purchased by the United States in 1917, the

Hexafoil symbol   WIK

spectacular islands had evolved into a tourist destination, offering to travelers some of the most beautiful scenery on the planet. However, a few entrepreneurial guides would capitalize on darker aspects, leading "ghost tours" that tempted visitors to connect with the spirits of those long-dead first occupants, slaves, and pirates. On this island—with its imaginary ghosts—Ed's journey would end.

Although St. John is as different as one could possibly imagine from Noxubee County far away in Mississippi, in terms of Ed's intentions here, there were some parallels. On this island wonderland, he was purchasing property to cultivate an exclusive place where only the most fortunate could live or visit. Legal in all aspects, the subdivision would result in expensive houses erected on separate lots, enhancing (or deflowering) the otherwise untouched terrain of this lush, undisturbed part of the island. It would certainly provide investments and needed construction employment. Yet not unlike the situation for the impoverished potential toxic-waste workers in rural Mississippi, the local workers would toil here in much the same ways, providing essential services with little promise of ever moving into the boss's house or his subdivision. Opposition also emerged from many who, like those in faraway Giles and Noxubee counties, were fearful of long-term environmental damage. No toxic waste machine was proposed, but the development and its economic profit would similarly remain unavailable for most of the population, many with ancestry dating back to the original inhabitants or imported slaves. Once again, wealthy people from elsewhere would win the game.

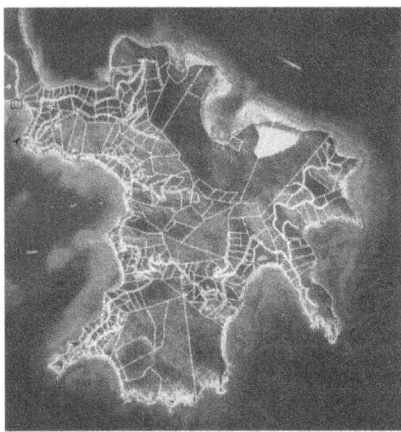
Properties at East End USVI

Ed's project would, unfortunately, continue the economic

disparity that already existed in this place where nature was at her most glorious. But that condition was not of his making and had been in place for many centuries, and it would continue to exist after Ed. Development of beautiful places can be profitable and, like the famous Seaside town in the panhandle of Florida, deliver a splendid quality of living for those who could afford it, and here he might have accomplished just that. He wouldn't be the first and definitely not the last to do so.

While Ed's subdivision project was being tilled on St. John, back in Tennessee, an encounter with one of my neighbors was bizarre and telling. My wife and I live on acreage at the back of a modest subdivision in Williamson County, our property for her horses protected forever with a conservation easement through The Land Trust for Tennessee. One day, the neighbor living next door shared with my wife that her daughter was dating a young man from Murfreesboro. Knowing that this was my hometown, the mom wondered if either of us knew him or his wealthy father. The boy was one of Ed's sons. Our neighbor described the lavish situations her daughter experienced with this handsome suitor, including visits to Ed's large farm and paid trips to the Virgin Islands where they stayed in a luxurious house on St. John. She further explained that Ed was impressed with her daughter and provided her with expensive gifts, confirming his rumored generosity. The girl's mother imagined that Ed was a tycoon, but she worried about her daughter's mention of hired guards providing protection, and that this precaution may have been taken due to common crooks on the island but perhaps also due to past or present adversaries.

My wife told the mother that both of us had known Ed many years earlier, and yet we knew nothing of this high level of living.

ISLAND OF PARADISE

We assumed that he was now even more successful than before, and that he now flew to paradise and lived like a sun-tanned baron. Later, the mom informed us that her daughter's relationship with Ed's son had ended. That was the last mention of my former classmate and adversary before I learned of his demise in that same Caribbean utopia of the East End.

Tropical beauty of East End   *SSP*

CHAPTER 27
The Final Pitch

In early 2014, Ed made his last known large insurance proposal to the Pasco County school district in sunny Florida. Located on the west central coast and primarily a bedroom community for nearby Tampa and St. Petersburg, Pasco County had an enormous number of public places of learning for its more than 60,000 students, and the county's educational district was ripe fruit for skillful insurance salesmen. The international reinsurance powerhouse Swiss Re and local management entity Pollock Financial Group formulated an attractive proposal, and due to his legendary, persuasive capabilities, Ed was hired as a gunslinging advisor and promoter. The insurance collaboration presented its scheme to the district for life insurance on the county's 10,000 teachers and staff members in a revised, private/governmental insurance program. Ed helped explain a fascinating concept to county officials in his convincing manner. This plan was similar to the Tennessee GOLI, where both survivors and the government would be beneficiaries. In the arrangement, the insured employee's family and the school board would receive

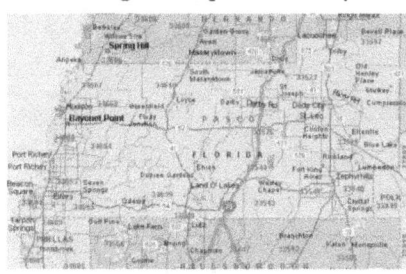
Pasco County, Florida   WIK

THE FINAL PITCH

the same amount of $50,000. However, it had a mystifying twist, for in this proposal, the district would pay nothing for the policies. Instead, four wealthy—and anonymous—New York families would each contribute $100 million to the plan. The combined $400 million pot of gold would be invested, its profits paying for the employee insurance policies. The distant investors would benefit from a tax shelter allowed on life insurance plans and beneficiaries. It was not a new concept and had apparently worked before, and here it was posed as merely a procedural exercise in which faraway wealthy Yankees would be helping teachers in the very deepest South. How nice it seemed.

The Swiss Re and Pollock promoters for this unusual perk for the county were hesitant to share details of the scheme, explaining that the specifics were proprietary and thus confidential. Ed and others assisting in the pitch were trained professionals at being coy, and they dodged critical questions on how and where the investments would be placed. It was also unclear what would happen if these

> ...ployees. This proposal would enable the board to pay $50,000 to an employee, and the board ...ceive the same amount, tax-free. Ed Netherland, a well-respected executive and insurance expe... ashville, guided the board through the most unusual idea. The most interesting (or rather perplexi... ea is that the premiums for the life insurance policies would be paid *not* by the school board but ... milies in New York, and in return these undisclosed families would benefit from tax-free returns o... it would pay the premiums. The process, a contract with the school board, would much needed f... ool board while providing a valuable "...

Coverage of school board meeting

funding techniques—essential to pay the insurance premiums in the coming decades—withered. The proposal was seen by elected officials, board members, legal advisors, and school employees as either a wonderful no-brainer or, instead, a convenient tax scheme benefiting four secret families living a thousand miles away. Based on the alarming results of an investigation of the plan's representatives by Florida regulators and stinging unsolicited criticism by national analyst Joseph Belth, the proposal was finally rejected by the county as too hard to understand, too risky, and probably too good to be true. Ed's last throw didn't hit the target, a rarity for the extremely successful pitcher of insurance. Within a few months, he would be struck out in a different way.

CHAPTER 28

Ending in Paradise

When the news broke in *The Tennessean* and other local newspapers about Ed's murder, several included an image of the crime scene, one that showed a modest yet architecturally stunning, cliff-edge cantilevered resort house overlooking the pristine waters of Sir Francis Drake Channel. It was a scene that could have been from the 1980s television series Fantasy Island. This was indeed a beautiful, idyllic setting for anyone desiring privacy with unbelievable nature surrounding in all directions. It would be difficult to find another place that more clearly defined the phrase "island paradise."

Ed had reportedly resided several times at other places on St. John such as the exclusive Peter Bay community. Consequently,

Newspaper coverage   *SJT*

there were questions as to why he was instead living at the time of his death in a relatively small, 800-square-foot getaway, one situated at the end of the most distant asphalt road on the nine-mile-long island. It was near the property that he had earlier hoped to develop, but he had reportedly lost ownership of those valuable acres in 2013, and now he was a solitary tenant in an expensive rental getaway. He had his trusty Jeep, a vehicle that he had customized and extended with a hauling bed and tailgate, serving not only as an agile four-wheel-drive road warrior but also as a truck. The landlord noted that Ed was always outside on his cellphone, and he was occasionally tardy in paying the monthly rent of $5,000. He found this tenant sophisticated and charming, and he knew that a check would eventually arrive. If Ed had wanted or needed to get away from it all, this was as far as he could possibly go.

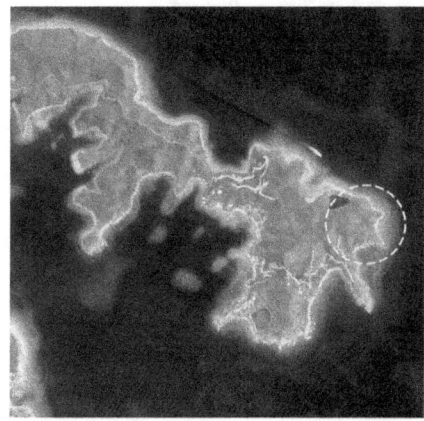
Aerial view of East End   WIK-N

Theories immediately abounded from his friends, business associates, and others regarding the reasons behind Ed's whereabouts in November 2014. One posited that he was hiding at this distant location in the rugged undeveloped property between Newfound and East End Bays, staying out of the public eye for soured business reasons. This theory is validated by a short email Ed sent to a close friend who might invest in the always-scheming creative thinker's "latest innovation." His message to another friend, sent in May 2014, simply said, "I finally got smart and stayed in the shadows." Another idea was that Ed sought some degree of safety in this hard-to-reach spot due to the island's notorious crime rate against the wealthy, a condition rarely publicized due to the territory's need for deep-pocketed visitors and investors. Tourists Kimberly Lapsey and James Malfetti had been murdered earlier in 2014, and both crimes were unsolved. Later, I learned from another associate of Ed that he did not feel particularly worried or threatened and may have simply been laying low due to complicated business and contentious legal reasons.

A third speculation was that the secluded, florally lush surroundings provided the privacy and quiet inspiration Ed sought while using email and his phone to easily conduct business in an uninterrupted manner. Due to his prior ownership of land near this isolated house, he knew the area well. He had reportedly received bypass heart surgery, so the peace, quiet, and ideal weather here would be nothing but totally soothing. In this bucolic setting of Oz-like beauty, he had an excellent northward view across the channel of a Caribbean-style Emerald City, one not of glass but of the majestic Mount Sage on nearby Tortola, largest of the British Virgin Islands. Handwritten notes found after his death confirmed that Ed was constantly developing another money-making idea, another insurance creation, or yet another way to return as the triumphant, brilliant strategist he had always been. It was the perfect, off-the-grid secluded outpost to formulate new battle plans. Or so it seemed.

The fourth and most bizarre explanation for Ed's situation possessed a frightening international flavor, one with the scent of a fictional thriller. According to a few friends, Ed was not in hiding but simply in need of financial recovery while he was trying to establish yet another money-making endeavor, and this time he had obtained the backing of a prosperous individual with rumored roots across the Atlantic Ocean. Ed may have convinced this individual to sink an enormous amount into one of his new projects, but it failed to sprout, and the large debt was making the investor mad and intolerant. It didn't help that Ed may have shown interest in an attractive Eastern European lady connected to this individual. He reportedly liked to drive fast, and he may have had a wild side, one reflecting his business bravado and temperament. Ed may have not known he was a revenge target, yet did this driver's patience run out of gas? If Ed could have been successful one more time using his acknowledged ingenuity and magician's ability to obtain cash fast, he might have pulled a financial rabbit out of his hat, and he might still be alive.

Meanwhile, on November 10, 2014, yet another legal entanglement was happening in Cincinnati, Ohio, where Fifth Third Bank had its national headquarters. It was related to the insurance lawsuits back in 2010 and 2012. According to a report by secondary market advisor Drinker Biddle, the financial giant was seeking "indemnification and reimbursement...for losses exceeding $100 million...allegedly sustained by the Bank as a result of the dishonest and fraudulent acts of the Bank's former employee, Matthew Ross...based on the interactions and activity of Ross and Ed Netherland of InsCap Management LLC." The report restated that Ross had colluded with InsCap officials Ira Brody, Harish Raghavan, and Ed to defraud the bank, and now the injured bank was going after all participating insurance carriers including Ace, Continental, Axis, Federal, and underwriters with famous Lloyd's of London. Only eight days later in the warm tropics, InsCap's main guy, who may have already paid dearly for the earlier financial wrath, would receive the death sentence from a ruthless criminal. The many contentious lawsuits with Ed's and other names attached would drag on for years with only slight notice of his murder.

ENDING IN PARADISE

To my knowledge, there were never accusations of illegal activities, drugs, or money-laundering in Ed's endeavors. Nor is there evidence that he inflicted serious physical harm to anyone. Many of his associates called him the greatest salesman they had ever known. Others said he was an incredible liar, had anger management issues, and could never be trusted. He may have produced hot air, but this ability had served him well in convincing individuals and companies to participate in his many schemes. It is probable that his legal and financial scorching of lenders and corporate investors was the main source of his problems and eventual downfall. Regarding anything related to the strange world of insurance, Ed had apparently twisted and stretched regulations almost to the breaking point. This is what many aggressive, and extremely successful ones often do. Was sheer ambition the cause of his death?

It ended for Ed on Monday night, November 17, or in the early hours of Tuesday, November 18, 2014. A cleaning service woman discovered his dead body that morning in the otherwise deserted house. News accounts reported that the sixty-year-old victim had received a fatal blow to the head, either from a weapon or due to a fall. It may have simply been a robbery gone wrong; another report suggested that the actual cause of death may have been traumatic heart failure. Had he known or followed local customs, he might have discouraged a local petty criminal with superstitious deterrents such as the hexafoil symbol or an odorous wreath. With only his wallet and watch taken from the crime scene, one could speculate that the killing might have been perpetrated by either a hired hand or a revenge-driven enemy. A scuffle was suspected, and there may have been more than one person involved in the act. Perhaps it was a fellow business associate or a female friend. Did the assailant drive quietly up the long road to the secluded place to do the deed? Or did the murderer(s) arrive by boat or kayak, carefully and silently scaling up the 100-foot slope from the beach to the house and then returning to quietly float away on the gentle waters?

Secluded getaway   SSP

Regarding the stolen watch, it was probably a lavish time-keeping device worth grabbing. Ed had always shmoozed with highly successful and rich fellas who owned only the best of anything and everything, and he clearly understood that to run in their leagues, he needed that symbol of status. To display his success, Ed probably wore a Rolex, Corum, or Patek Philippe to visually convey a cultured taste for the finer things. However, for the practical mind, it seemed ostentatious to pay for these outrageously expensive watches, frequently appearing in magazines and on billboards on the wrist of a gorgeous actress, handsome actor, or famous athlete with the not-so-subtle visual message that their stardom was due not only to their stunning looks and achievements but also to their exquisite, ticking piece of jewelry. One could always be on time with a more workaday timepiece, but such an inexpensive tick-tock object would not have impressed those in Ed's universe, nor, on that terrible night, his assailant(s). Ed's stolen watch was probably not a Timex.

The St. John police said they performed an investigation and an autopsy, soon announcing the cause of Ed's death in an antiseptic manner: "blunt-force trauma." It was a local crime, yet the victim had been a jet-setter with known business and personal connections on the island and in Florida, Tennessee, and with folks across the United States and overseas. Assistance from the FBI on this United States property did not immediately materialize to the great frustration of Ed's friends who lived on the island and many of his relatives back home. This agency would have certainly been able to do a more thorough job than the local force with DNA analysis, and they could have checked airlines and hotels for any and all known to have associated with Ed or his companies in the Virgin Islands and throughout the United States. If it was a Mafia-style "hit," the likely motive for the killing—normally involving money, pride, or both—might have been discovered. A lead might surface either from evidence found in the thick underbrush or nearby calm waters of this location, or from the many complicated personal and business relationships that Ed had cultivated, ones that might have required him to seek the sadly insufficient protection of this outpost. The police chief ruled that this case was simply a robbery with a fatal ending. But many other items of value were reportedly in Ed's possession at the remote cottage, and these weren't stolen. Why? Unfortunately, one could also not dismiss the possibility that some island officials wanted this difficult-to-solve-crime to just fade away, the investigation placed on ice and forever frozen on this perpetually hot, Caribbean rock.

CHAPTER 29
Funeral Service

Ed's body was returned to Murfreesboro, and his funeral was scheduled for November 26. It would be held under the auspices of the same Church of Christ denomination for which his father had preached years ago. Online comments to the funeral home managing his burial were warm, with a few kind expressions from fellow students at Bellwood Elementary School and Central High. Because I had an obligation in Murfreesboro in the late afternoon of that day—and due admittedly to sheer curiosity regarding what had occurred in paradise more than a week earlier—I decided to include the visitation in my agenda. My goal was to stay for only a moment, speak with others who would show up, and then discreetly leave with those who would not or could not stay for the service. A minuscule possibility was that my high school flame—Netherland's temptation target long ago—would also appear to show her last respects.

On the way to the church on this temperate November day, I stopped first at nearby Evergreen Cemetery to visit the graves of my parents. They had been laid to rest a few years earlier in a newer, flat section of this spacious, historic cemetery, and I frequently stopped by to check on their headstones, replace artificial flowers, and pause in permanent grief, one softened only by immense gratitude for

FUNERAL SERVICE

my upbringing in this wonderful college town. My late father had taught music education at the nearby university, and my late mother was a teacher at Bellwood Elementary. Ed may have been one of her students, but I don't recall her mentioning him. As I departed my parents' resting places and headed to the early-afternoon visitation, I noticed in the distance of the vast, flat cemetery, an open-sided green tent covering a freshly-dug grave, and I wondered if this was the one prepared for Ed's final address. It was.

I walked into the church foyer just as the funeral service for Ed was about to begin, and I hugged several coeds from our days at Central High. I signed the register and glanced through the pages to see who else had shown up at the previous evening visitation or earlier in the day. I recognized a few names. Glancing around at others in the narthex, I gladly shook hands with my old friend Hollywood, who was now a director with the nearby funeral home handling the burial arrangements. Because there had been many funerals of friends' parents in my hometown during the past years, I had seen him in this situation all too often. Today, he was again in charge but this time for a man that he and I had grown up with. He had been friends with Ed during Central days and also at UT where he played on the varsity team that his classmate had helped manage and train. The two connected only occasionally after college, and my former country club worker and high school and college classmate was here to comfort the family and provide the needed assistance in his always professional manner.

I peered inside the large sanctuary, then decided to step inside for a brief moment, immediately greeted with the strong, distinct scent not of lilacs but lilies. The service was being preceded with a continuous series of large projected images from Ed's life, primarily with his children. Unable to resist seeing this unfamiliar side of this person from the past, I was urged by a dear classmate and life-long friend Vicki to stay. I could not resist, so I sat down beside her on the very back pew at the center aisle near the main sanctuary doorway. Happy times were shown on the large screen. He certainly loved his kids, and they him. I did not see his second wife and later learned that they had divorced. In looking at the congregation of mourners, I noticed that Hollywood and I were apparently the only males in attendance from our large graduating class back in 1972. This was awkward and hard to understand. I was poised to leave, but the female friend reminded me that my afternoon appointment was still a few hours away. She said, "You're not in a rush, so stay here with me and others." Seated on that back pew, I was positioned beside the main doors that provided a convenient exit hatch. At any point in the upcoming commemorative event, I could easily and quietly depart.

The funeral service commenced, and soon I heard a quite different tale about this murdered classmate and father. The two ministers and others eulogized a very special person, with heartfelt praise for Ed's lifelong devotion to his kids, his remarkable inventive financial deals, his generous charity, and his loyal friendship to others. He was a Chartered Life Underwriter and a member of the Million Dollar Round Table, the premiere association of financial professionals. He ran with the best in his field and savored the tasty flavors of impressive insurance successes. Naturally, there was no mention of his nefarious activities and the numerous scandals, yet with all the references to humility, Christian devotion, and late-in-life spiritual recommitment, I was somewhat dumbfounded. This certainly wasn't the person I had remembered. To them, Ed

FUNERAL SERVICE

lived life in its fullest manner, and those associated with him—yet never burned—recalled his endearing swagger. I also learned of his lifelong friendship with the football star Terry Sneed, the most formidable member of Central's 1970 state champion football squad. Sneed was perhaps the greatest football player to ever come through the Murfreesboro school system, and after he died unexpectedly a few months earlier on June 4, Ed provided emotional and financial support to his family. Though he was apparently not engaged or active in community or civic organizations, charities, or nonprofit organizations, Ed clearly had sides and aspects I never knew. Was he benevolent and generous to many unknown others? In an article penned shortly afterwards, local newspaper columnist Sam Stockard commented that he wished he could have located more coverage of Ed's good works.

CHS star Terry Sneed

CHAPTER 30
Aftermath

During the immediate months after Ed's murder, I found very few who expressed complete surprise at his mysterious death. Many thought it was more than a botched robbery. Given their memory of him and with revealed, publicized coverage of his lifelong strategy and tactics, controversial projects, strained relationships, and numerous lawsuits, the prevailing opinion of many was that he had finally crossed someone who decided to resolve their differences not in a courtroom but in a more painful, physical manner. A few speculated that Ed had finally done something so shady, financially ruinous, or insulting that he was killed in a savage act of revenge. Past classmates and I wondered about his children and how this tragedy would forever shape their lives, hoping that their futures would include good memories of a father who attempted and often succeeded where others would not and could not.

Why had a private moment with Ed been on my bucket list? Long ago, this classmate had cheated me and I had never forgotten it. I remember his upstart actions and treatment of others years ago in high school. Unsatisfied grudges are very lightweight and can, until resolved, rest quietly and patiently on one's shoulder like a gentle yet toxic flower. Ed wanted to steal my girl, and I despised him not only for his sneaky actions but equally for the pompous way he tempted her in my presence. I had hoped to someday settle the score and felt justified to confront him and attempt to knock his haughty smirk into the next hemisphere.

Ed probably couldn't have cared less as I was never any threat to him. Long ago, he got that girl, and he moved on in the pursuit of other relationships and enormous wealth. Little had I known of his ingenious, sometimes unsavory and contorted business and personal activities in the following decades. Later I would learn that in the waiting line of past and present adversaries demanding resolution, I would have been at the very rear. I discovered activities that suggested the strong probability of a confrontation with, or retaliation by, a very determined and deadly opponent. Our talents and interests had always been different and totally unrelated, and except for his early Casanova triumphs, I had no envy of him, and he likely never had envy of me. I had simply remained silently furious over that act of overt temptation long ago and his frequent unkind treatment of others, and I wanted to deliver an appropriate repayment. The encounter with Ed might have been a verbal skewering, a fistfight (which I might lose), a standoff, a truce, or perhaps the lighting of a peace pipe. He could have thrown a compliment, noting that we both had excellent taste in pretty ladies, but he would likely strike with a barb of bitter reality, reminding me that my attractive sweetheart

Ballon wine, vintage 1973

had not discouraged his advances, and he certainly wasn't going to ignore the potential delicious nectar. Our meeting might have even ended with a handshake and a smile, the two of us reconciled and sharing our separate stories as we raised a glass of balloon wine made in 1973 to that country girl of long ago. I'll never know.

Criminal charges for Ed's murder were never made. Six months after the incident, his frustrated oncologist Robert Oldham complained directly to the distant local St. John Tradewinds publication that the homicide investigation by the local police seemed lackluster, questioning why the FBI was not invited to help solve the case. He noted that given the sluggish, barely crawling pace of the investigation, someone was going to get away with murder. This sentiment was echoed by the lenient, rent-forgiving landlord of the house where Ed was killed.

Though not in the initial news coverage, it was later revealed that blood-stained sheets had been discovered on the shoreline below the vacation cottage, adding a new wrinkle to the crime and possibly validating the theory of an attack by water from the nearby lovely inlet of Newfound Bay. It was also rumored that Ed's body had been thoroughly pummeled, and this suggested a retaliatory nature of the assault. In the early part of 2016, the local island police force again explained that the case was "still under investigation." From discussions with a local newspaper reporter, I learned that FBI agents eventually looked into the murder and were pursuing leads, but they did not release any information, encouraging or not. My attempt to obtain the latest official news through the FBI's Freedom of Information process eventually resulted in a lackluster, standard response, and the agency disappointingly closed the case.

The final nails were probably driven into the investigation's

AFTERMATH

coffin during the fall of 2017 by the devastating one-two punch of hurricanes Irma and Maria. Both of these dreaded category-five storms thundered through St. John and the neighboring islands, and what Hurricane Irma didn't destroy with its record-breaking winds, Hurricane Maria drowned just two weeks later. With such devastation, there was an immediate focus on recovery and repair. The unsolved murder of a single person with a complicated, questionable past three years earlier now seemed irrelevant, and hopes for any further police investigation were blown away. Although the remote house where Ed was killed had survived both hurricanes, whatever traces of evidence from the crime remaining on the grounds

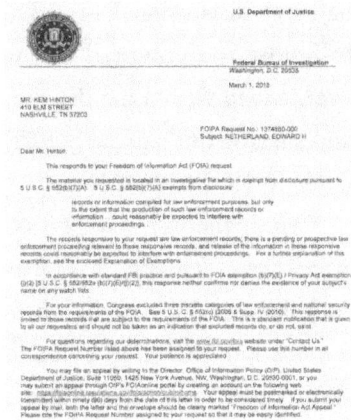

Letter from FBI

Hurricane Irma

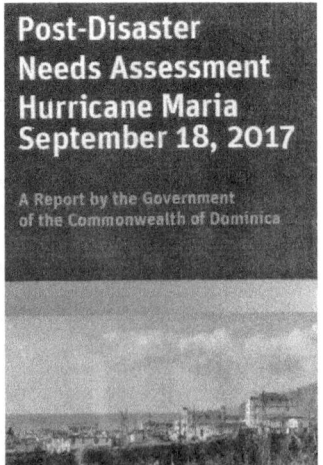

WIK Hurricane Maria

were probably forever washed—or cleansed—away.

Another, unverified and perhaps never verifiable tale was that a diabolical, life-threatening warning had been delivered perhaps a year or less before Ed's murder to one of his close friends or an employee in the Boro. From a very unhappy investor, the message was to firmly convey that money lost was not money forgiven. This seemed to confirm one hunch that the borrowed money may have been gone and there could be no repayment. Perhaps retribution of a most unkind physical manner was issued by this investor or by his/her henchmen.

CHAPTER 31
The Mirror

My issue, now seemingly ridiculous, was with Ed and not his family. I remember his older sister at Central High was the distinguished head majorette of the Tiger marching band, but I never knew the other siblings. I assume their loss is immense, especially because they don't know who committed the crime. Until solved, I assume there is only so much that could or now can be done to ease that perpetual remorse. Too many are lost in the unfair, dark fog of never knowing. I hope his innocent kids receive his love of family and his knowledge, confidence, initiative, and stamina. Someday, perhaps they and his siblings, relatives, and friends will learn of the truth. His killer remains unknown and free, and the unsolved case on that scorched island gets colder each year. Ed's oncologist and close friend Robert Oldham and I spoke about their relationship, and he remains furious after a decade that someone is out there with blood on their hands, a criminal who took away someone he admired and cared for. Ed deserves justice, and his family and authentic friends like Oldham deserve closure.

Because I wanted feedback and verification of Ed's many endeavors, insurance successes, and related activities, I circulated different drafts of this memoir over the past several years to friends, a few former classmates, and those who might provide additional, appropriate information. Exceptional advice on these initial texts was received from skilled authors, notably Keel Hunt, Rick Glaze, and the late Robert Hicks. Famed U.S. attorney Hal Hardin stressed the importance of accuracy, that the truth is always essential. These initial versions unfairly possessed disdain for Ed, for I had received many testaments of his dark side. An unhappy investor considered him ruthless, someone who cared only about his children and never worried about those he damaged. "Ed could be crazy and erratic," stated a former employee, adding, "He didn't seem to be in touch with the reality of his obligations. He'd be overly demanding and often mean to his staff. I got fed up." Another in the Boro was simply amazed at how Ed had the hubris to promote money-making yet unwanted projects against justified opponents, ones like those citizens in Giles and Noxubee counties who were defending their cherished lands from something toxic and totally terrifying. Puzzled, she wondered, "Did he have no regard for their real fears and of their hatred of his sheer existence?"

However, when I shared a version with one of his financially injured associates, she was only full of remorse for his surviving kids. She expressed no desire for revenge but, instead, only total forgiveness, and she was not pleased with my words, thinking my views and those from others were lopsided. I needed that slap on the face, and I realized that the ancient grudge may have poisoned me with some sense of justified retribution. Was Ed an authentic character with twin faces and contradicting values, our little town's strange case of Dr. Jekyll and Mr. Hyde?

Other different stories of Ed surfaced. An adorable majorette from our days at Central High wanted me to know that Ed earned her gratitude by guarding her reputation from the false words of an obnoxious, bragging football player. My discussion with oncologist Oldham had also delivered a valuable and distinctive point of view that held significant value and impact. Ed's engaging Boro banker also provided this assessment: "The guy was bombastic, extraordinarily smart, and loyal. He was also frustrating, because he seldom took my advice. Yes, he had adversaries, but it never slowed him down, and he lived life to the fullest. We were very close, and I considered Ed a younger brother." One of Ed's many attorneys added, "He never took no for an answer. He was persuasive and persistent."

The incomplete drafts about Ed were also the subject of discreet discussions at the Central High School class of 1972's fiftieth reunion in the early summer of 2022. Of the nearly 500 graduates on the huge, now-faded composite photo, more than 90 showed up, a decent number given the sobering half century of elapsed time. Several brought their yearbooks from that now-distant era, opening pages to recall and share the images of favored (and hated) teachers and the touching messages from fellow classmates now more than 18,000 days older. Most possessed a sincere joy to see how their fellow students had maneuvered through life, and yet

Central High 50th booklet

the reunion's commemorative booklet, printed and distributed a few months later, would sadly remind attendees that more than sixty of their fellow classmates, including Ed, were forever gone. Because the other Ed named Eddie's body was still missing, he wasn't listed.

If asked about Ed's murder and the disappearance of that other Ed, I was willing to share what I knew and had discovered, and I observed whispers among those who remembered both of them. Most thought that the famous boy Ed had simply become fabulously rich and was then abruptly robbed and killed, while the vanishing of the more reserved Eddie was more sinister. Few if any knew of brilliant Ed's contentious incinerator proposals or of his unique, frankly astonishing insurance and investment inventions. Those that knew both Ed and me from CHS days back then simply wondered why I was interested in dead Ed, why I even cared. I didn't share that it was due to something inexplicable and eerie that I had experienced at the end of his funeral eight years earlier.

Central High 50th reunion   RB

The reunion also delivered a special confirmation of the wisdom of graduating seniors long ago when all of us had voted for special citations bestowed to the best fellow students, awards called "superlatives." The individuals selected fifty years earlier for Most Outstanding, Most Likely to Succeed, Most Intellectual, Most Popular, and other recognitions did indeed fulfill their potentials, and seeing many of them at the gathering proved delightful and quite satisfying. Perhaps we were actually gifted at that young age in knowing who would rise to the top and honor that final class of Central High. Ed had not been in that select group, but he certainly should have received recognition, one in a category not bestowed that year, as Most Ambitious.

A full decade after his murder, Ed's name and mysterious demise are still mentioned. Our surviving group of Baby Boomers from Central High have crossed that unavoidable birthday line numbered seventy, and if we believe Mark Twain's definition of the worst kind of lie, statistics, we know that we are likely in our last decade. That challenging reality causes reflection, and I constantly wonder why my life has been the way it is, why something at the end of Ed's funeral happened, why I needed to know the totality of his rise and fall, and how my opinion of him is now a complicated, conflicting bouquet of confusion, respect, fascination, and, for his family, sadness. What would he have done if not in that spectacular place at the most wrong time?

I have always been fascinated, admittedly infatuated, with events and the people of my days in high school and college, cherishing many of the memories and intoxicated by moments long past. Now more than ten years after the murder of Ed in a distant place, this saga about him is perhaps a more balanced tale of one who had been injected with an extremely effective steroidal dose of sheer, unadulterated ambition. Both of us had privileged upbringings and lifetime fortunes, and I have attempted to show his. Mine has included guidance and total devotion from my late parents, completion of college and graduate degrees, the enduring love of a faithful woman, acquisition of land for her horses, the raising of a son who obtained a PhD, a fascinating career in architecture, survival of emergency brain concussion surgery, recovery from heart surgery to fix a heart murmur before it would permanently fix me, and the fortune of having so many authentic friends, especially from my formative years, that they would overfill UT's enormous Neyland Stadium.

The ups and downs in my life on the gentle, rolling hills of Middle Tennessee pale when compared to Ed's frequent wild, abrasive, thrilling, occasionally foolhardy, and triumphal scaling of financial heights like Mount Everest, too often followed by a tumble into one of the nearby deep icy crevasses. He is gone; I am not. I have many obligations still to fulfill including recognizing the humanity and value of everyone in my past and present life. A rebuttal to what I have composed cannot come from Ed but might arrive from others in his life. Ed's survivors may not appreciate my side of his story, and they might be justified. I had opened a chest full of shimmering insurance gems, treasures of business gold, black lumps of legal coal, and the glittering orb of a mysterious death. I saw and heard what I didn't see or hear before. In reality, both of us were privileged white Boomers, and perhaps we shared far many more similarities than differences. I will always wonder what was on Ed's bucket list.

The remote modern house on St. John where Ed met his final moments was refurbished. It is again available as a private getaway featuring extraordinary views of the ocean below and of majestic Tortola Island directly north. It was and is an excellent

Beautiful cottage   SSP

candidate for a personal retreat and, perhaps one day, the subject of a captivating postcard. A restful, reflective visit to this beautiful cantilevered place of solitude is now on my amended list of places to visit. While there, would the ghost of the brilliant insurance executive from faraway Tennessee revisit the striking, contemporary architectural structure? He and all other patient spirits in the occasionally dark skies overhead might just drop down and drop in. Their visits to this and other appropriate spots across the island would allow them to share the intriguing and complicated history of a spectacular place and the experiences of prior inhabitants of St. John, good and bad, recent and long ago.

CHAPTER 32
Haze of Purple

Hollywood asked for my help as Ed's funeral ended, and I stepped up from the back pew. The minister had told the congregation that the depositing of the casket down into the soil of Murfreesboro was to be family-only private. The cherry wood container was rolled slowly down the center aisle. My friend motioned me to walk ahead and hold open the numerous doors leading to the arrival area where a hearse was parked and silently waiting. Once outside, the back door of Ed's final taxi was opened, yet there were no designated pallbearers, only Hollywood, myself, and a few young men, probably nephews, standing beside the casket. In silence, we together lifted the heavy box and placed it on the extended metal rails and then guided it back into the black vehicle. As it moved inward, I placed my right palm onto the smooth surface of the casket containing this nemesis from my past. It was a strange, numb feeling, with no sense of hate, retaliation, sorrow, triumph, or farewell. Nothing.

Honoring the minister's earlier request, I did not engage with the surrounding family and others nearby. I had no business being there, yet was. Beneath the neutral sky, one in which ancient spirits from faraway places and different eras may have assembled to observe, I slowly walked to my car. My eyes did not leave the hearse and the silent box within. Then my glance turned upward, and a

HAZE OF PURPLE

distinctive haze approached. I immediately recognized its delightful scent, an intoxicating fragrance made from crushed floral petals. For a few seconds, I was completely enveloped in that purple mist, hearing the sound of the color, and not knowing if I was coming up or down. From above, the faint voice from a Stratocaster-playing phantom, gone since 1970, suggested that I should kiss the sky.

Then, all around, the soft cloud evaporated. It was a spiritual and supernatural encounter, and soon I might know why.

In that moment, I could not comprehend what had happened nor conceive of the source of the lavender fog. I drove home in a stunned trance, traveling in a white Porsche 911 that I had wanted four decades earlier. Completely forgetting the scheduled afternoon appointment, I belatedly realized that I would have been absolutely worthless at that meeting or at any other personal interaction. Except for the purring of the sportscar's low RPMs and the noise of a void, I heard nothing during the slow and purposely longer-than-necessary trip on rarely-driven country roads as I passed through the counties of Rutherford, Davidson, and, finally, to my place in Williamson. Only afterwards would I learn of the enormity of Ed's many financial escapades and lifetime adventures. Pallbearer duty had been mistakenly assigned to me, and the mystifying shroud of that regal color may have been a message. As this person faded away, did a clear challenge step forth? Was it a purgatorial request or a demanding dare? It left me with

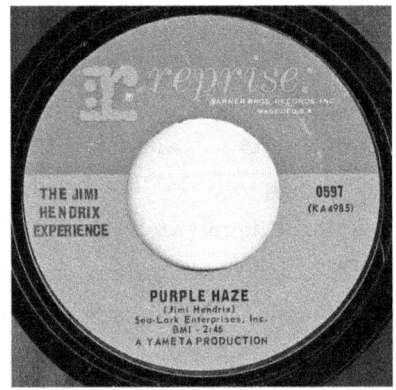

"Purple Haze" by Jimi Hendrix

an obligation to discover Ed's days after UT and his "pedal-to-the-metal" business excursions. This would be joined with an increasingly revved up curiosity to know why the brakes on his wild trek had been cruelly stomped.

Rest in peace

One small item on my bucket list will never be achieved. Instead, its possibility was stolen by the actions of an unknown, ruthless criminal in a setting of immense natural beauty. It had become impossible to achieve in this life, and now due to my increased knowledge of a cruelly murdered classmate, the item was gently blown away to be a note on a different ledger, one of possible rapprochement. The eventual answer about the strange life of Ed Netherland, a complicated person of immense aspirations, bullish confidence, unforgiving abrasiveness, unapologetic adventuring, ingenious inventiveness, and (almost) perpetual maneuverability—yet one with mysterious scruples and spun from a perplexing type of moral cloth—will have to wait for a time and place, either high above or in the nether far below, where the two of us will meet again.

CREDITS

Graphic Design: Jim Vienneau of VNO Design
Editing: Katy Yocom, Sam Severn, Ashley Hagan, Billie Chrisman, and Michelle Adkerson

I am most grateful for encouragement and advice from: Keel Hunt, Hal Hardin, Rick Glaze, Cecil Elrod, Billie Chrisman, Katy Yocom, Robin Hood, Jim Vienneau and the late Robert Hicks.

Special appreciation is extended to: Colin Crawford, author of *Uproar at Dancing Rabbit Creek: Battling Over Race, Class, and the Environment* (published by Addison Wesley Publishing Company, 1996) for his exhaustive research on Ed Netherland's attempts to establish toxic waste facilities in Giles County, Tennessee, and in Noxubee County, Mississippi. His 410-page publication documents the economic injustices that had existed and continue to exist in too many places in the United States. Crawford is now Dean of the Bowen School of Law at the University of Arkansas. I thoroughly recommend his book to all Americans.

Additional gratitude to: Ronnie Barrett, Karen Byrd Bauer, Gary Brewer, Susan Findley Brown, Allison Bulpitt, Jane Carroll, Julie Christie, Vicki Anderson Christopher, Laura Deleot, Martha Sykes Felker, Duane Fouts, Janet Basse Graham, John Mack Green,

Lindsay Hagar, Trent Hanner, Andy Hayes, Lillian Hibbett, Marilyn Mabry Hinton, Garry Hood, Kay Horner, Kayla Howard, George Huddleston Jr., Pallie Walker Jones, Bill Ketron, Christine Kreyling, John Lancaster, Chuck Lewis, Susan Loyd, Bracken Mayo Sr., Jim McCawley, Rick McKnight, Andrew McMahan, Buddy Meeks, Buddy Miller, Hugh Moffatt, Jennifer Howard Murphy, Jeana Nunley, Tom Oat, Randy O'Brien, Robert Oldham, Beth Lokey O'Leary, Robert Owen, Alex Palmer, David Parsons, Billy Pittard, Jayme Proctor, Susan Huntington Rothchild, Anita Haynes Stanley, Scott Stewart, Melissa Stroop, Debi Taylor Tate, Susan and Royce Taylor, Tori Thomas, Ralph Turley, Phil Willis, Mitch Wilson, and Beth Morgan Wright.

INFORMATION SOURCES

Sources of information include: *Chicago Sun Times, Chicago Tribune, The Insurance Forum, Investment News, Knoxville News Sentinel, Knoxville Journal, Macon Beacon,* Murfreesboro's *Daily News Journal, Murfreesboro Post, Murfreesboro Pulse, Nashville Post, The New York Times, News of St. John, Pasco Times, Pasco Tribune, The Pulaski Citizen, St. John Source, St. John Tradewinds, Tampa Bay Times, Tampa Tribune, The Tennessean,* University of Tennessee's *Daily Beacon, Virgin Islands Daily News;* interviews with individuals in Murfreesboro, Pulaski, Macon, Nashville, Knoxville, New York, Jacksonville, St. John Island USVI, and other locations; public records of Rutherford County Library System Historical Research Center; public records of Metropolitan Nashville & Davidson County; archives of Albert Gore Research Center at Middle Tennessee State University; collections of the Giles County Historical Society; public records of Chicago, Cincinnati, Jacksonville, Tampa, Pasco County, New York City, St. Thomas Island USVI, St. John Island USVI; courtesy of the Tennessee State Library & Archives; archives of the University of Tennessee; responses from United States Federal Bureau of Investigation; photographic websites Wikimedia Commons, Freepik, and Unsplash; and *Uproar at Dancing Rabbit Creek* by Colin Crawford (Addison-Wesley, 1996).

PHOTOGRAPHS

With exceptions as noted, photographs are by the author, from entities below, or from individuals, or from historic and contemporary images available on the worldwide web.

AWP	Addison-Wesley Publishing Company, Inc.
ADG	Anderson Design Group, art by Joel Anderson
COL	Colorama
FRE	Freepik
IMA	IMAGO/Panthermedia
IMG	Imagn Images: Karen Kraft-*USA Today Network*
KNS	*Knoxville News Sentinel*
MB	*Macon Beacon*
MT	MTSU Albert Gore Center
NSJ	*News of St. John*
PEX	Pexels: Betul Kara
RB	Ronnie Barrett Photography
RCH	Rutherford County Historical Research Center
RDS	Reeves Drug Store, Pulaski
SJT	*St. John Tradewinds*
SME	Sony Music Entertainment
SSP	Steve Simonsen Photography
TPC	*The Pulaski Citizen*
TSLA	Tennessee State Library & Archives
TSLA/MG	TSLA/My Genealogy
UNS	Unsplash: Amanda Horitz
USVI	USVI Planning Department
UT	University of Tennessee Archives
WIK	Wikimedia Commons
WIK-N	Wikimedia: NASA
WIK-P	Wikimedia: Mike Peel
ZAZ	Zazzle

CENTRAL HIGH CLASS OF 1972

AUTHOR BIOGRAPHY

Kem Hinton is an architect, urban designer, visual artist, historian, and author. He was born in Nashville, raised in nearby Murfreesboro, and holds degrees from the University of Tennessee, University of Pennsylvania, and the Ecole des Beaux-Arts in Paris. He was founding partner with Seab Tuck of Tuck-Hinton Architects (1984-2019). Their firm's architectural work received state, regional, and national awards, and their buildings have been publicized in prestigious international design journals of Italy, France, Germany, and Japan. Kem is a Fellow of the American Institute of Architects, served on the Tennessee Historical Commission (2020-2024), and recipient of the AIA Tennessee "William Strickland Lifetime Achievement Award." He is now studio director of Kem Hinton Design.

The nonfiction memoir *Nether Land* is Kem's fourth book. He is also the author of *A Long Path: The Search for a Bicentennial Landmark* (Hillsboro Press, 1997), *Seablets* (Mascot Books, 2019), and *Tennessee's Bicentennial Mall* (Grandin Hood Publishers, 2022).

www.ingramcontent.com/pod-product-compliance
Lightning Source LLC
LaVergne TN
LVHW020829271125
826378LV00008B/157